# Security Analytics

## Chapman & Hall/CRC Cyber-Physical Systems

Series Editors:

**Jyotir Moy Chatterjee**
*Lord Buddha Education Foundation, Kathmandu, Nepal*

**Vishal Jain**
*Sharda University, Greater Noida, India*

**Cyber-Physical Systems: A Comprehensive Guide**
*By: Nonita Sharma, L K Awasthi, Monika Mangla, K P Sharma, Rohit Kumar*

**Introduction to the Cyber Ranges**
*By: Bishwajeet Pandey, Shabeer Ahmad*

**Security Analytics: A Data Centric Approach to Information Security**
*By: Mehak Khurana, Shilpa Mahajan*

For more information on this series please visit: https://www.routledge.com/Chapman--HallCRC-Cyber-Physical-Systems/book-series/CHCPS?pd=published,forthcoming&pg=1&pp=12&so=pub&view=list?pd=published,forthcoming&pg=1&pp=12&so=pub&view=list

# Security Analytics

## A Data Centric Approach to Information Security

Edited by
**Mehak Khurana**
**Shilpa Mahajan**

CRC Press
Taylor & Francis Group
Boca Raton London New York

CRC Press is an imprint of the
Taylor & Francis Group, an **informa** business

A CHAPMAN & HALL BOOK

First edition published 2022
by CRC Press
6000 Broken Sound Parkway NW, Suite 300, Boca Raton, FL 33487-2742

and by CRC Press
4 Park Square, Milton Park, Abingdon, Oxon, OX14 4RN

*CRC Press is an imprint of Taylor & Francis Group, LLC*

*Library of Congress Cataloging-in-Publication Data*
Names: Khurana, Mehak, editor. | Mahajan, Shilpa, editor.
Title: Security analytics : a data centric approach to information security
/ edited by Mehak Khurana, Shilpa Mahajan.
Description: First edition. | Boca Raton : Chapman & Hall/CRC Press, 2022.
| Series: Chapman & Hall/CRC cyber-physical systems | Includes
bibliographical references and index. |
Identifiers: LCCN 2021060075 (print) | LCCN 2021060076 (ebook) | ISBN
9781032072418 (hardback) | ISBN 9781032265261 (paperback) | ISBN
9781003206088 (ebook)
Subjects: LCSH: Computer security--Data processing. | Computer
networks--Security measures--Data processing. | Data mining. | Machine
learning--Industrial applications.
Classification: LCC QA76.9.A25 S37566 2022 (print) | LCC QA76.9.A25
(ebook) | DDC 005.8--dc23/eng/20220322
LC record available at https://lccn.loc.gov/2021060075
LC ebook record available at https://lccn.loc.gov/2021060076

ISBN: 978-1-032-07241-8 (hbk)
ISBN: 978-1-032-26526-1 (pbk)
ISBN: 978-1-003-20608-8 (ebk)

DOI: 10.1201/9781003206088

Typeset in Palatino
by SPi Technologies India Pvt Ltd (Straive)

# Contents

# Preface

The aim of *Security Analytics: A Data-Centric Approach to Information Security* is to encourage the community of multinational researchers to showcase the research work done in their field of security analytics. This medium provided an opportunity to researchers on an international forum to learn about the latest developments through scientific information interchange in the field of Cyber Security, Cyber Physical Systems, and Analytics.

This book was conceived after analyzing the increase in the prevalence of cybercrime attacks on business organizations, government infrastructures, and individuals.

It focuses on analysis of data based on context, value and compliance controls surrounding the data being secured. This view helps us apply appropriate technical and local security controls depending on what is being secured. Data-centric security may result in reduced security controls surrounding less critical systems and data, saving on resource usage and budget, which can be applied to improving security in more sensitive areas. The content promotes a multidisciplinary approach that reflects the requirement of cybersecurity, security analytics and forensics in various other domains.

The topics were categorized, namely: Cyber Physical Systems; Cyber Security; Blockchain; Network Security; Mobile Security; Security in IoT; Cloud Security, Web and Mobile Security; Security in Data Analytics; Security Analysis Using Machine Learning, Cryptology, and its applications; Cyber and Digital Forensics; Network and Mobile Security, Blockchain, and Software Technologies. We received a good number of submissions all over India and overseas; each submission was anonymously reviewed by three reviewers. After extensive reviews and shepherding, 13 chapters were accepted, and this book includes revised versions of all accepted papers. We must mention that the selection of the papers was an extremely challenging task. We would like to thank everyone who contributed directly or indirectly in making this book a success and ensured its smooth running. The support of technical partners is also appreciatively acknowledged. **Dr. Mehak Khurana and Dr. Shilpa Mahajan** have coordinated the whole process of promoting, editing, and compiling the book.

# *Editors*

**Mehak Khurana** has more than 11 years of experience in teaching and research. She is currently working as an Assistant Professor at The NorthCap University, Gurugram. Prior to that, she has worked at HMRITM, GGSIPU. She earned a Ph.D. degree in the field of Information Security and Cryptography. She completed her M. Tech degree with a silver medal in Information Technology from USICT, GGSIPU, Delhi. To better align the department with the industry's best practices, she introduced and designed courses on Penetration Testing, Secure Coding and Software Vulnerabilities, Web and Mobile Security. She also organized International Conference on Cyber Security and Digital Forensics in 2021 in collaboration with Springer as a Convener. Her specialization is cybersecurity, information security, and cryptography. She has published many papers in various reputed National/International Journals and conferences. She has been a mentor to various B. Tech and M. Tech projects. She has been a resource person in various FDPs, workshops, guest lectures, and chaired the session at various Conferences. She is editing books from the various publishers like Springer, CRC Press, etc. She is serving as a reviewer for many reputed Journals and TPC member for various International Conferences.

She has also been involved in activities to improve the standard of the profession. For that, she has coordinated many events in collaboration with IIT Bombay, IIT Madras, and IIIT Delhi. She successfully organized TEDx 2017 and 2019 as a co-curator. She is leading a few chapters and societies in her current university, Institute of Engineers India (IEI) CSE chapter, Programmers Club, Alumni Sub-committee, and Open Web Application Software Project at the North Cap University, and many more.

She is an active member of various global societies, such as the Open Web Application Software Project (OWASP), Optical Society of America (OSA), Women in Appsec (WIA), Cryptology Research Society of India (CRSI).

**Dr. Shilpa Mahajan** has more than 14 years of teaching experience at postgraduate and undergraduate levels. She is a committed researcher in the field of sensor networks and has done her Ph.D. in the area of Wireless Sensor Network at Guru Nanak Dev University, Amritsar. She completed her post-graduation with distinction from Punjab Engineering College, Chandigarh. She specializes in Cyber Security, Computer Networks, Data Structures, Operating Systems, and Mobile Computing. She has introduced and designed various courses like Network Security and Cyber Security. Presently two doctoral scholars are pursuing their Ph.D. under her supervision. She has guided various M. Tech and B. Tech Projects. She has published many research papers in peer-reviewed reputed international journals and conferences. She has been a resource person in various FDPs, workshops, guest lectures, and seminars.

She is a CCNA certified instructor and has also done certifications in Data Scientist Tools, Exploratory Data Analysis, and Getting and Cleaning Data from Johns Hopkins University. She is a Lifetime member of ISTE. She is a CISCO-certified training instructor for CCNA modules 1, 2, 3, and 4. She has been appointed Advanced Level Instructor this year. She received an appreciation from Cisco Networking Academy for 5 years' active participation. She also set up a CISCO Networking Academy and developed a CISCO lab at NCU, Gurgaon in January 2014.

# Contributors

**Gaurav Aggarwal**
Department of IT and Engineering
Amity University Tashkent
Uzbekistan, Central Asian

**Aparna Bannore**
Department of Computer Engineering
SIES Graduate School of Technology
Mumbai, India

**Ardhani Satya Narayana Chakravarthy**
Department of Computer Science and
    Engineering
JNTUK-University College of Engineering
Kakinada, India

**A. Charan Kumari**
Department of Computer Science
Dayalbagh Educational Institute
Agra, India

**Nitin Chhimwal**
Department of Computer Science &
    Engineering
Birla Institute of Applied Sciences
Bhimtal, Uttarakhand, India

**K. Ganga Devi**
Department of Computer Science
Vels Institute of Science and Technology
    Advanced Studies (VISTAS)
Chennai, India

**R. Renuga Devi**
Department of Computer Science
Vels Institute of Science and Technology
    Advanced Studies (VISTAS)
Chennai, India

**R. Girija**
Department of Computer Science
School of Computer Science and
    Engineering
VIT University
Chennai, India

**S. L. Jayalakshmi**
School of Computer Science and
    Engineering
VIT University
Chennai, India

**Keshav Kaushik**
School of Computer Science
University of Petroleum and Energy
    Studies
Dehradun, Uttarakhand, India

**Nirmal Kirola**
Department of Computer Science &
    Engineering
Birla Institute of Applied Sciences
Bhimtal, Uttarakhand, India

**Priyanka Maan**
Department of Computer Science and
    Engineering
The NorthCap University
Gurugram, India

**Perattur Nagabushanam**
Department of EEE
Karunya Institute of Technology and
    Sciences
CBE, India

**Kabeer Nautiyal**
School of Computing and Information
    Technology
Manipal University Jaipur
Jaipur, India

**Rachana Y. Patil**
Department of Computer Engineering
Pimpri Chinchwad College of
    Engineering
Pune, India

**Seyedali Pourmoafi**
Department of Computer Science and
    Engineering
University of Hertfordshire
England, United Kingdom

**Subramanyam Radha**
Department of ECE
Karunya Institute of Technology and
    Sciences
CBE, India

**Nagarajan Ramalingam**
Department of Electrical and Electronics
    Engineering
Gnanamani College of Technology
Namakkal, India

**Bidar Sachin**
Department of Computer Science and
    Engineering
University of Hertfordshire
England, United Kingdom

**Bhavna Saini**
School of Computing and Information
    Technology
Manipal University Jaipur
Jaipur, India

**Sudhir Sharma**
Manipal University Jaipur
Rajasthan, India

**Hukum Singh**
Department of Applied Sciences
The NorthCap University
Gurugram, India

**Shyamal Srivastava**
Department of Computer Science
School of Computer Science and Engineering
VIT University
Chennai, India

**Kannadhasan Suriyan**
Department of Electronics and
    Communication Engineering
Cheran College of Engineering
Karur, India

**Sandesh Tripathi**
Department of Computer Science &
    Engineering
Birla Institute of Applied Sciences
Bhimtal, Uttarakhand, India

**N. Umamaheshwari**
Department of Computer Science
Vels Institute of Science and Technology
    Advanced Studies (VISTAS)
Chennai, India

**Anju Yadav**
School of Computing and Information
    Technology
Manipal University Jaipur
Jaipur, India

**R. Vedhapriyavadhana**
School of Computer Science and
    Engineering
VIT University
Chennai, Inida

**Seema Verma**
Department of Computer Science
Delhi Technical Campus, Greater Noida
Faridabad, India

**Manas Kumar Yogi**
Computer Science and Engineering
Pragati Engineering College
    (Autonomous),
Surampalem, India

# 1

## A Reliable Blockchain Application for Music in a Decentralized Network

**Bhavna Saini**
*Manipal University Jaipur, Jaipur, India*

**Gaurav Aggarwal**
*Amity University in Tashkent, Tashkent, Uzbekistan*

**Anju Yadav and Kabeer Nautiyal**
*Manipal University Jaipur, Jaipur, India*

## CONTENTS

## 1.1 Introduction

Blockchain technology is frequently called the new internet that can advance the form of various enterprises. The music business is no particular exception in that regard and is important for artists who want to get paid for their creative work. Blockchain vows to give authority back to content creators by providing a proper distribution channel. Here, the concept gives a platform where they will have the freedom to get valued data and be paid rightfully [1].

### 1.1.1 Need of Change in Music Industry

The emergence of the internet and music streaming platforms has changed the industry altogether. As a result, the value chain restructuring has accelerated the rise of various music streaming platforms [2].

Figure 1.1 portrays how the shipment of CDs has dropped over time whereas, on the other hand, Figure 1.2 illustrates the increase in the revenue from online streaming platforms over the last decade [3].

This change affects everybody in the music business, such as artists, labels, distributors, lyricists, and streaming service providers. The strategy by which music eminences are determined has consistently been a tangled one. Nonetheless, the ascent of internet use has made it significantly more intricate.

The unique limitations and prerequisites of the modern age music industry represent some significant difficulties to accomplish a straightforward procedure. One of the hurdles is the heterogeneity of the included actors, partners, and the existing management models. It also includes various degrees of privacy, thus leading to the absence of transparency [3].

Figure 1.3 clarifies the progression of the artist's licensed innovation with every one of the players associated with the creative process prior to distribution.

- **Artist:** Provides creative content to the music industry.
- **Recording Engineer:** Handles the technical aspect of the recording process.
- **Record Label:** Provides different marketing campaigns and promotional strategies for the creative work of artists. They are responsible for the final products, such as music albums.
- **Distributors:** Provides the market to the creatives. They are responsible for the sale of albums.

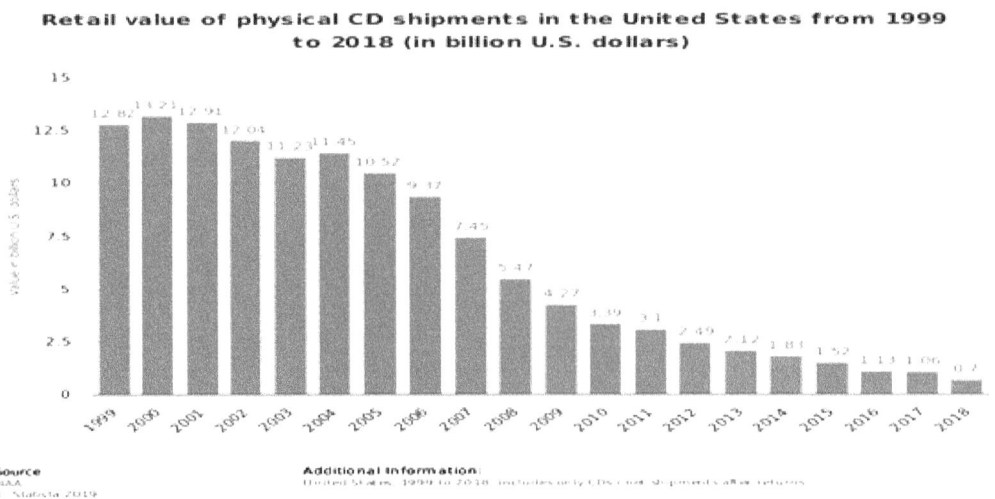

**FIGURE 1.1**
Retail value of CD shipment in the US [3].

**Revenue from music streaming in the United States from 2010 to 2018 (in billion U.S. dollars)**

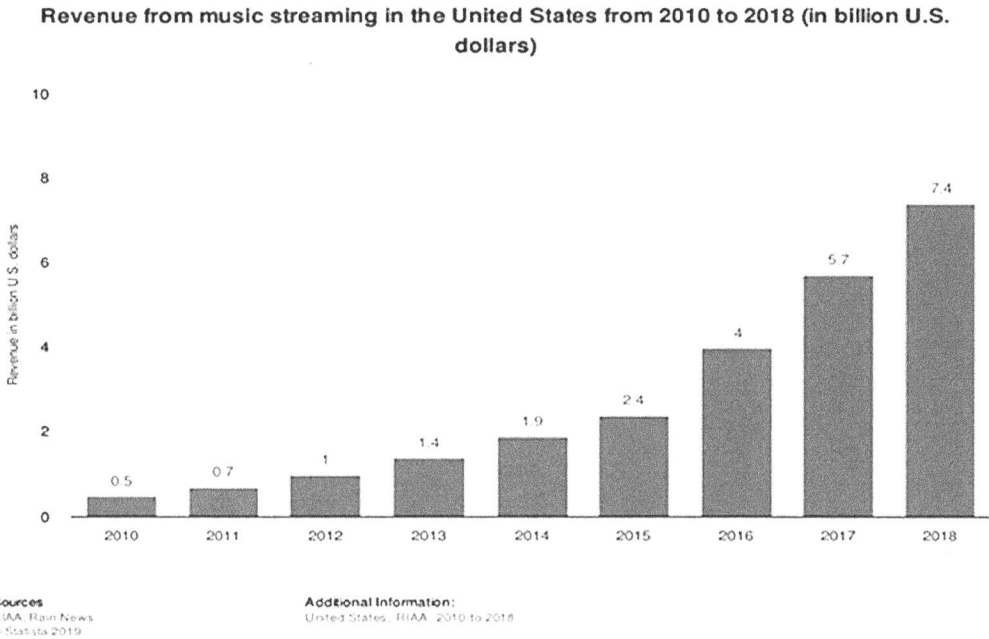

**FIGURE 1.2**
Revenue from music streaming platforms[3].

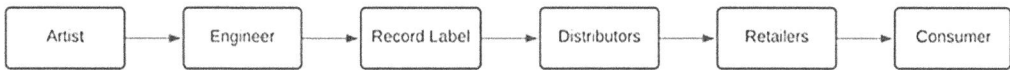

**FIGURE 1.3**
Intellectual property flow.

- **Retailers:** They are the ones who sell the physical copies of the creative work of the artists. Different streaming platforms can also be considered modern-day retailers.
- **Consumers:** The final element of the chain.

The music business is driven by purposeful misdirection – merchants and record labels are generally not ready to reveal who claims the rights to what music, in what region, and what sort of utilization. In the interim, fundamental data inform deals between content makers and mediators (for example, marks, distributors, and spilling administrations). Figure 1.4 shows the traditional flow of royalty payments. In this antiquated framework, the middle people take value-based charges with no practical present-day support, and royalty payments are frequently postponed [4].

## 1.1.2 Challenges Faced by the Music Industry

Because of an absence of straightforwardness, creatives working in the business face plenty of issues. There are three central points of interest:

A) Firstly, there is no unified information base that archives the ownership rights to a melody and connecting them to a specific artist. All things being equal, there are

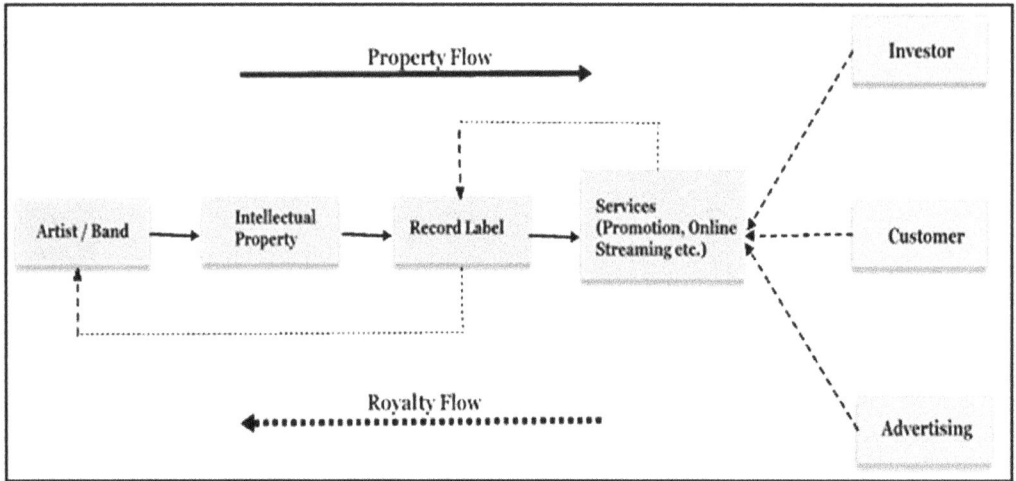

**FIGURE 1.4**
Traditional royalty flow.

different data sets, none altogether far-reaching. Fundamentally, when the work is co-possessed data might vary, starting with one information base then onto the next, with no focal power to resolve any struggles.

B) Secondly, payments are slow, and the distribution of the royalty is not transparent. Particularly for international owners, it takes years for royalties to reach the bank accounts of the rightful owners.

C) The third issue is that impressive sums are regularly paid to some unacceptable party and critical assets of sovereignty income end up outside the artist's scope. Here the meriting proprietors of sovereignty income really shouldn't still be up in the air as a result of an absence of an industry-wide framework [5].

D) The fourth issue is the absence of transparency in the value chain.

This digital platform gives rise to transparency in royalty distribution and payment [6]. This is where the blockchain will come into the picture. With many users on the internet, blockchain technology will help maintain a distributed music database with proper ownership in a public ledger. In addition to this, smart contracts will be used for a royalty payment, maintaining transparency and fairness.

## 1.2 Introduction to Blockchain

Blockchain can be illustrated as a data structure that holds value-based records, ensures security, straightforwardness, and decentralization. As shown in Figure 1.5, the data/information is recorded in an open ledger, recording every transaction taking place. Blockchain is a sequence of records or blocks that are linked together using cryptographic methods. A decentralized organization doesn't need any transitional element. The information about each trade anytime completed in the blockchain is shared and open to all hubs. This property makes the structure more straightforward than bringing together exchanges including

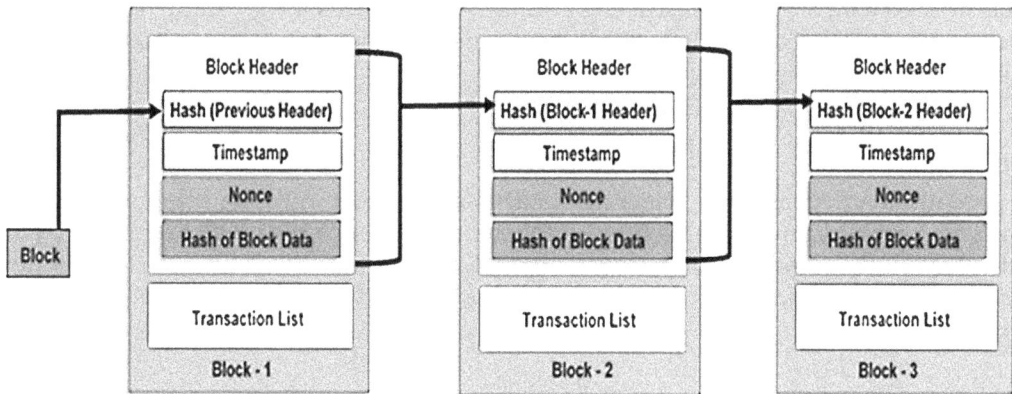

**FIGURE 1.5**
Blockchain technology.

an outsider [7]. For instance, consider that object A needs to send cash to protest B on the organization. To finish this exchange, an initial block is created which contains every bit of the information. This block is then communicated on the web for approval. At the point when approved, this block is added to the chain, and thus the conditional interaction is finished.

This innovation shines bright after the broad acknowledgment of Bitcoin, the absolute first digital cryptocurrency, has become one of the most discussed topics in today's techno-logical world [8]. Every exchange on a blockchain is secured with a digital signal that demonstrates its genuineness. Because of the utilization of encryption and computerized signatures, the information stored in the blockchain is carefully designed and can't be changed. In plain terms, Blockchain is a continuous database that keeps up with constantly growing data records that are affirmed by the nodes taking part in it. However, there is more to blockchains than just information security. As exemplified by Bitcoin, the essential feature of blockchain innovations is that the information itself can be decentralized. A broadly spread public record upheld by a blockchain where all users have similar informa-tion, which is significant for high-esteem use-cases, for instance, money, is obviously secu-rity intrusive for some, utilization cases. Information security and protection on the blockchain is a creating field [9].

## 1.3 Blockchain Technologies

### 1.3.1 Smart Contracts

One basic use instance of blockchain innovation incorporates "savvy contracts." A savvy contract is a convention proposed to carefully work with, affirm, or maintain the trade or execution of an understanding. Savvy contracts permit the execution of solid exchanges without outsiders. These exchanges are identifiable and unavoidable [10]. These agree-ments are programmed programs that are mechanized to execute the conditions of an arrangement. When a predefined condition for a brilliant agreement is met, the authoritative concurrence with a lawfully restricting arrangement can be made according to the conditions laid out [11]. Consequently, blockchain innovation is appropriate in both the monetary and non-monetary universes.

### 1.3.2 Ethereum

Ethereum is an open-source, blockchain-based circulated processing stage and system including shrewd agreement usefulness. Ethereum is quick to try a total execution of this idea. Ether is the cryptographic money made by Ethereum as a motivation to excavators for calculation performed and is the lone cash acknowledged in the installment of exchange expenses [12].

## 1.4 Framework for Royalty Distribution Using Blockchain

As seen in the previous sections of this chapter, there are many processes, legal bindings, and actors involved in the whole royalty distribution system followed currently. With so many intermediaries present, the process has lost its transparency and the deserving are not being paid for their work correctly. With the implementation of blockchain technology, one can solve many aspects of these problems and similar problems faced in the music industry. Figure 1.6 shows the framework for royalty payments using blockchain technology. It explains how blockchain will revolutionize the music streaming sector and benefit artists in every possible way.

The block content represents the beholder's permissions and data possession shared by members of peer-to-peer and private networks. Blockchain technology encourages the use of "smart contracts" on an Ethereum-based ledger [13], which allows us to automate and track specific state transitions (such as a change in viewership and ownership rights).

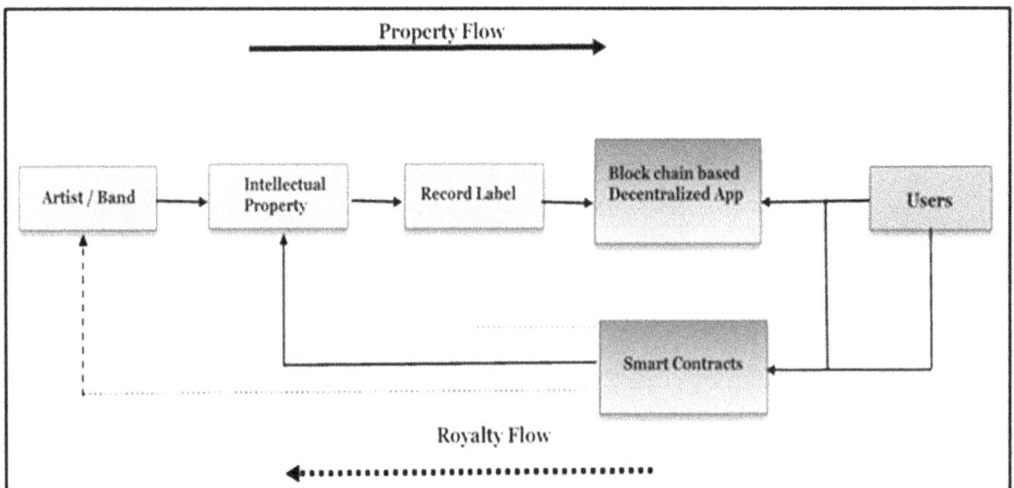

**FIGURE 1.6**
Royalty distribution using blockchain.

Via smart contracts based on Ethereum, a relationship can be established between a user and the artist, which associates a song with the viewing permissions and data retrieval instructions. A cryptographic hash has been included on the blockchain to ensure against tampering, which provides data completeness.

Artists can put their new songs out to the distributed database, which will be added to the ledger. Any uploads on the network by the artists will be tracked via their Ethereum ID. Users can access these songs through a smart contract. Based on the contractual agreement (royalties), when both parties agree mutually, users will be provided with access to the particular song. The artist has access to the updated database, which contains the purchaser's unique ID and location. This keeps the artists engaged with the growth of their records.

To navigate the potentially large number of users and artists on our network, our model has introduced four types of contracts for managerial ease.

A. Registration Contract

This agreement relates the member's distinguishing proof strings to their Ethereum address identity. Our model uses strings rather than alphanumeric public keys. This permits the usage of a generally existing type of ID. Arrangements and regulations coded into the agreement can enroll new users or change the planning of existing ones. Name enlistment would thus be able to be confined distinctly to guaranteed and enrolled names to prevent theft of any sort.

B. Streaming Contract

This contract is used to establish a relationship between the user (buyers) and the artist. If a user needs to stream a song or an album of a particular artist, they first need to agree to a smart contract that will be stating all the conditions stated by the artist and their management. When agreed upon these conditions, a specific amount is deducted from the user's wallet and distributed automatically between the label and the artists as per the agreement. As soon as the transaction is completed, the user's Ethereum ID is linked to the song's Ethereum ID, hence providing them with access to the song.

C. Download Contract

This contract is very similar to the streaming contract, mentioned in the second point of this section. The primary difference between a streaming contract and a download contract is, this contract is used when an end-user needs to download a song. To avoid multiple streaming costs, users are presented with an option of downloading a song, so that they can listen to it even when not connected to the network. An additional downloading cost is associated with this contract, which will be automatically deducted from the user's wallet.

D. Royalty Contract

This contract is explicitly used when a record label is playing a role in terms of the distribution of the songs. In such a case, a royalty contract is agreed upon by the artist and their management which records that the royalty distribution rates have been agreed upon mutually. The payment contract is drafted as per the royalty contract.

E. Share and Collect

Share and collect is a simple method by which a user or an artist can earn their cryptocurrency. This works in the same way a referral code works. For example, one can share a song with others and the more the people listen to this song through the unique code shared, the more cryptocurrency they earn. The duration of the listening is also taken into account. This will help the artists to grow more and have a wide listening audience.

F.  Miners

   Miners are the benefactors of their computational resources and members in the organization that add to accomplishing a dependable and slow headway of the chain. Our model depends on Ethereum's innate boosting where exchanges require an organization's money unit; for example, Ether, to be changed by the organization. Ether can be obtained by mining, tackling a computational riddle, and granting a sufficient measure of it to that hub.

Artists are in this manner urged to partake in mining to back the continuation of their exercises. In like manner, when users wish to access an artist's content, they will need to spend Ether. At the point when the block is mined, the capacity consequently abuts the block's miner as the abundance proprietor. The excavator can demand this abundance and gather it [14,15].

## 1.5  Implementation and Results

On analyzing the literature regarding blockchain in music, it was recognizable that the attention was mainly given to the "blockchain side" of it. This chapter aims to provide a superior image of the music business by recognizing all the actors associated with the recording industry. Only from that point onwards was it possible to outline the new blockchain models [16].

   Our model focuses on providing artists with their entire growth history and maintaining transparency. Blockchain technology refers to the identification of an actual copyright holder that can be quickly made. Also, it provides easiness in tracking the derivative works through the value chain.

   This section provides a series of images (see Figures 1.7–1.12), of how the app will be working for the end-user. In Figure 1.7, the end-user will be registering themselves on the network, where unique Ethereum ID will be linked to their account. A unique identification number is required, for example, a passport number, so that there are no duplicate accounts present on the network.

   Figure 1.8 depicts the primary interface of the decentralized app. A wallet can be observed in the top right corner, which keeps track of the cryptocurrency available in hand to the user.

   Figures 1.9–1.10 shows that whenever a song is streamed on the network, some amount is deducted from the user's wallet as soon as the user agrees to the smart contract. The cryptocurrency obtained from this transaction is distributed among the artists and their management automatically.

   In the event of a download, the user needs to toggle the download switch. As soon as the switch is toggled, a pop-up notification similar to that shown in Figure 1.9, is displayed on the screen. When agreed about the specified amount, a predefined smart contract will automatically deduct the amount from the end-user's wallet.

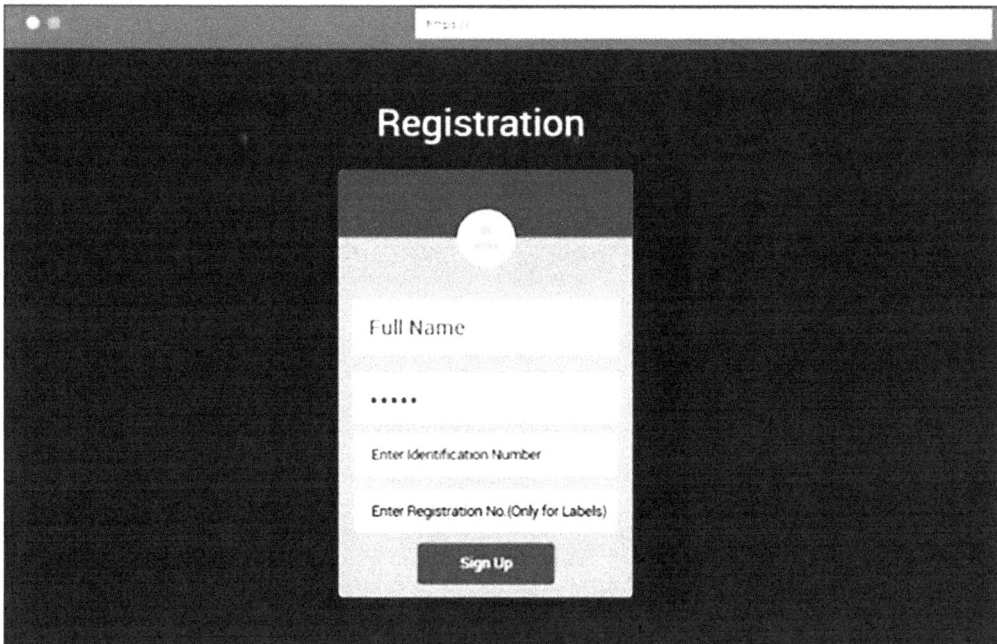

**FIGURE 1.7**
Registration form which will link the user to the unique Ethereum ID.

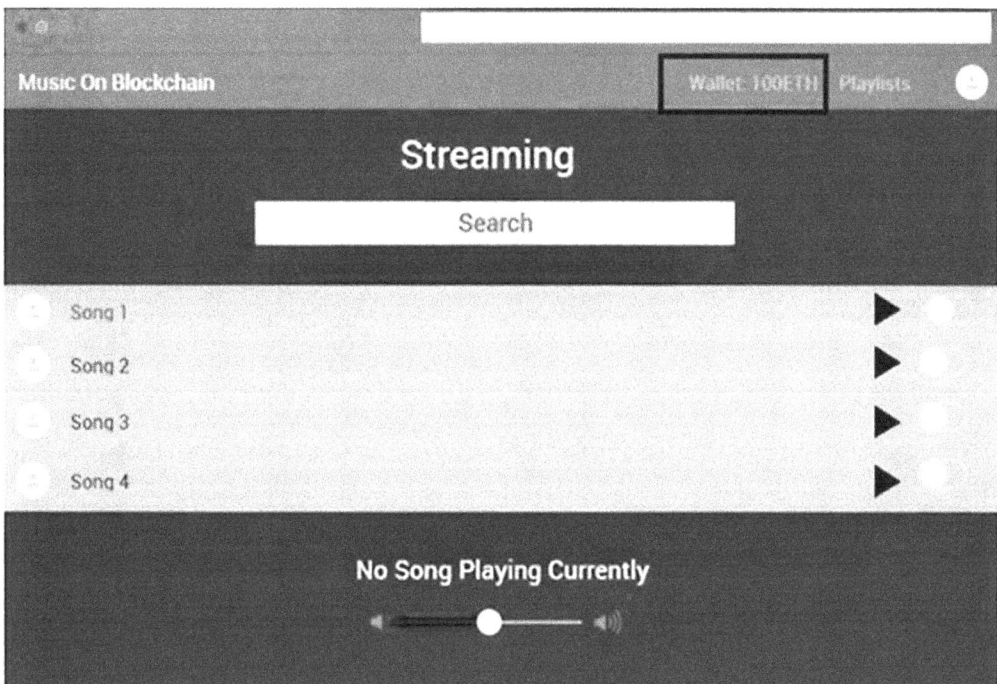

**FIGURE 1.8**
Shows the primary interface of the network.

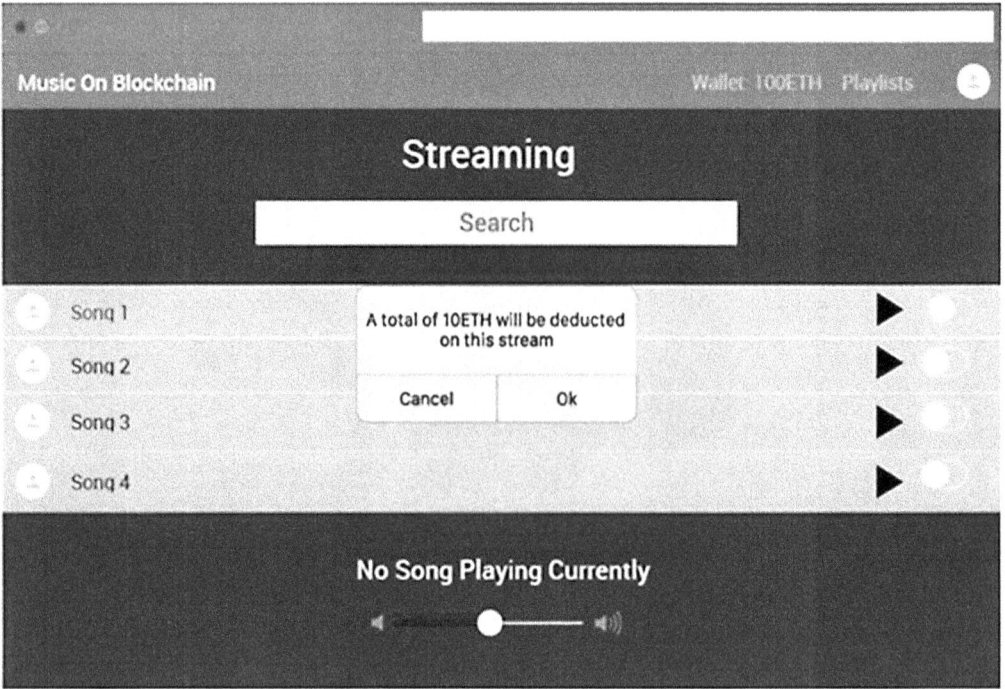

**FIGURE 1.9**
Smart Contract notification generated.

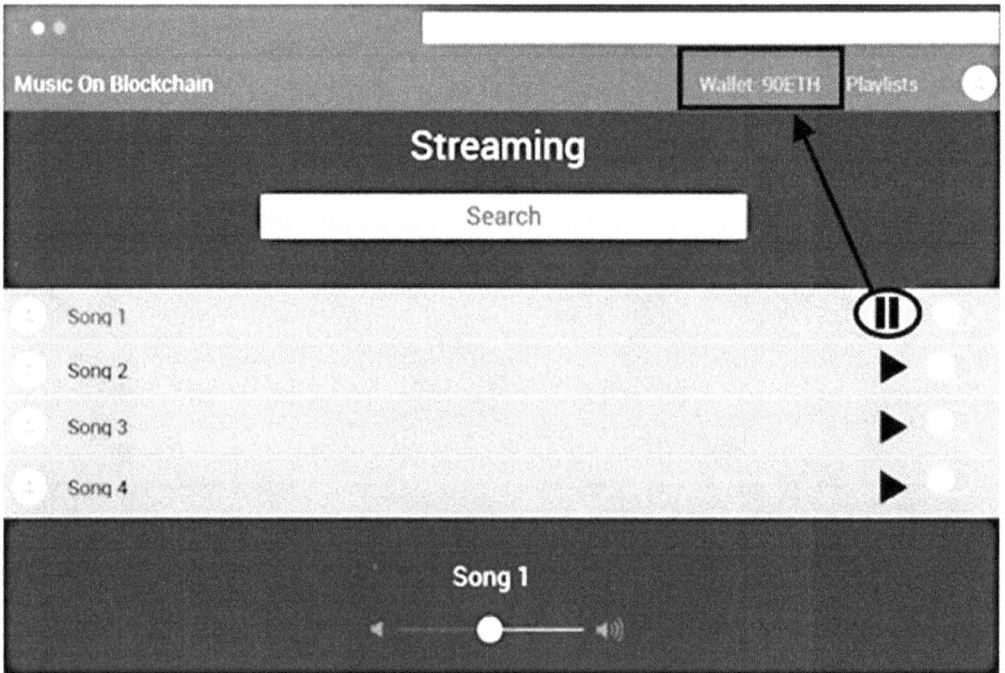

**FIGURE 1.10**
Deduction of the specified amount from the user's wallet.

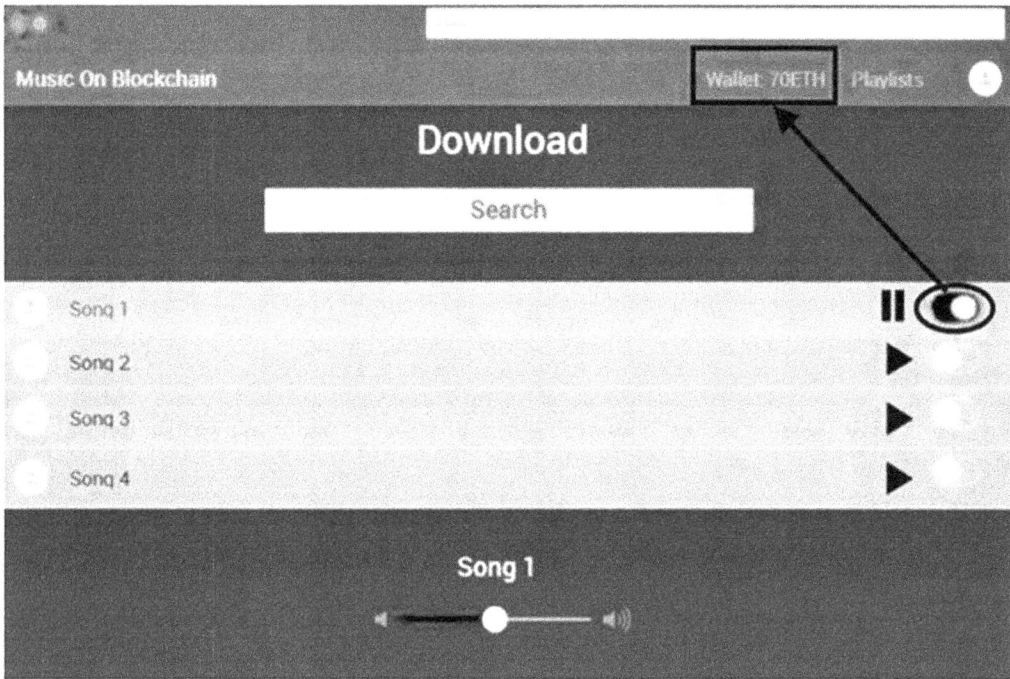

**FIGURE 1.11**
Amount deduction in event of a download.

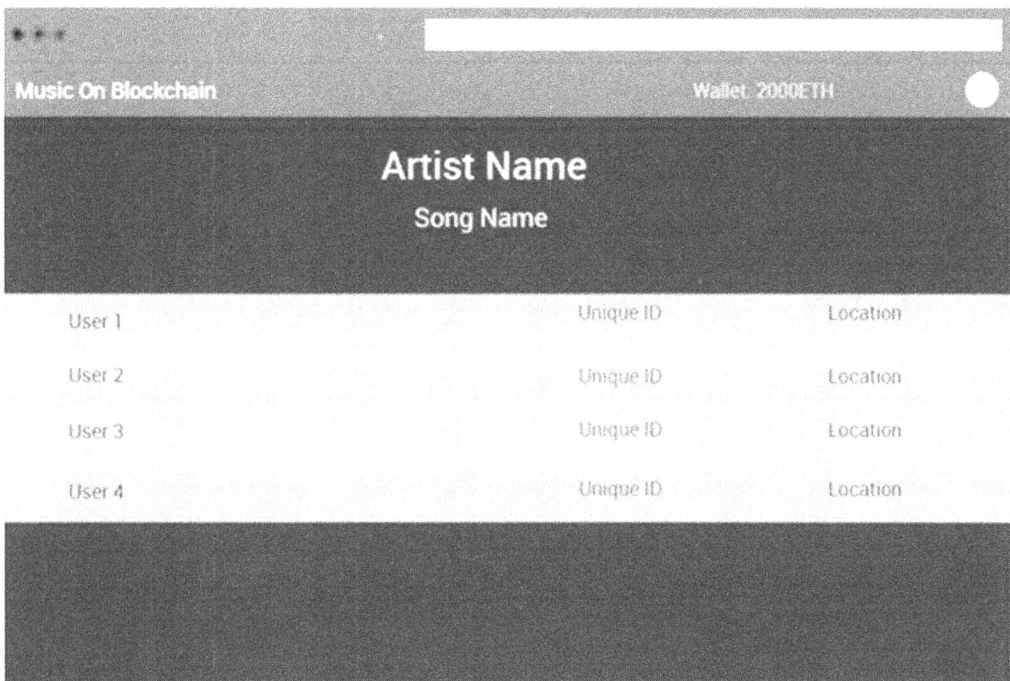

**FIGURE 1.12**
Artist page displaying information of all purchases.

Figure 1.12 shows the Artist Status page, which contains all the details of the transactions made on a particular song or an album. This is where transparency is maintained on a blockchain.

## 1.6 Future Research Prospects

In this study, it was evident how technological growth has impacted the music industry. The digitalization of music and its distribution has changed the industry altogether. As previously discussed in Section 1.3, the online streaming of music has increased dramatically in the past decade. The emergence of technologies such as blockchain into the music business is just a start toward a new generation of music streaming and distribution. The application of blockchain is restricted to fair royalty distribution, but it can be applied further.

### 1.6.1 Universal Music Database

As the copyright framework has been developed because of innovative changes in the last one hundred years, how the copyrights are abused falls under various sets of rules and regimes. In the digital marketplace, the administration of copyright enforcement on sound recording has become tricky. Artists and record labels battle to discover fair approaches to permit their work [17]. With further research in the application of blockchain in the music domain, one can overcome these problems.

### 1.6.2 Digital Content Distribution and Licensing

After the commercialization of the internet in 1994, digitized content has been expanding rapidly. Particularly music and video content would involve the more significant part of its traffic. There are two sorts of distribution frameworks; one is secured, and the other is unsecured. The procedure to ensure digitized content has a great deal of variation. In any case, the essential technology involved is encryption. With the high encryption power and a decentralized database, more research into blockchain's application can help us devise a mechanism that would solve all the existing issues [18]. Blockchain technology is based on decentralization and aims to have secure storage of recordings, distribution, and trade of digital works.

### 1.6.3 Affordable Music Sampling

Music sampling, also known as "digital sampling," uses components of earlier released songs inside another arrangement. As a result of the commercial success of selection and legal licensing, the sampling market has gotten progressively dynamic and active. Along with the explosive demand, copyright holders understood that permitting samples of their music could be a rewarding business. However, only big stars were getting to use this technology, whereas independent artists were left in the dirt. With the assistance of blockchain technology, the middle person can be removed. Thus, bringing down the barrier of access inside a typical framework, artists will be able to contact the owners of copyrighted work for licensing purposes directly. This will also reduce the piracy of music to some extent.

## 1.7 Drawbacks

An obvious constraint faced while studying the subject is the implementation of the whole situation (the innovation as well as its application to the industry), which leads to the absence of practical case studies and real-life scenarios. It is hard to state if the organizations wandering into this region will be successful.

Also taken into account are the scalability and the feasibility of the devised project. Coming from its name, blockchain represents a chain of records. This means that each new record is connected to the whole chain and each block becomes bigger and bigger. This leads to multiple issues, such as more time will be needed to process each transaction, bearing in mind that it also needs to be verified. More blocks will also affect the running costs.

## 1.8 Conclusion

In this chapter, the problem of unfair royalty distribution in the music industry has been discussed. Rectification of this problem statement with the help of a decentralized network, blockchain technology, and smart contracts has been attempted. Utilizing blockchain innovation, this work has shown how standards of decentralization can demonstrate support in data management of the music business at a huge scope. Along with this, maximum attention is given to maintaining transparency at the highest level. The further enhancement and development of blockchain in the music industry can be utilized by composers and artists, as well as consumers.

## References

1. Sitonio, C., & Nucciarelli, A. (2018). *The impact of blockchain on the music industry*. Available online: http://hdl.handle.net/10419/184968
2. Arcos, L. C. (2018). The blockchain technology on the music industry. *Brazilian Journal of Operations & Production Management, 15*(3), 439–443.
3. Rethink Music (2015). *Fair music: Transparency and payment flows in the music industry*. Berklee Institute of Creative Entrepreneurship, Boston, MA
4. Taghdiri, A. (2019). How blockchain technology can revolutionize the music industry. *The Harvard Journal of Sports and Entertainment Law, 10*, 173.
5. O'Dair, M., & Beaven, Z. (2017). The networked record industry: How blockchain technology could transform the record industry. *Strategic Change, 26*(5), 471–480.
6. Marques, M. C. M. (2019). Another technology transforming the Music Industry: Blockchain (Doctoral dissertation).
7. Xie, J., Tang, H., Huang, T., Yu, F. R., Xie, R., Liu, J., & Liu, Y. (2019). A survey of blockchain technology applied to smart cities: Research issues and challenges. *IEEE Communications Surveys & Tutorials, 21*(3), 2794–2830.
8. Chen, H., Pendleton, M., Njilla, L., & Xu, S. (2020). A survey on ethereum systems security: Vulnerabilities, attacks, and defenses. *ACM Computing Surveys (CSUR), 53*(3), 1–43.

9. Halpin, H., & Piekarska, M. (2017, April). Introduction to Security and Privacy on the Blockchain. In *2017 IEEE European Symposium on Security and Privacy Workshops (EuroS&PW)* (pp. 1–3). IEEE, Paris, France.

10. Hewa, T., Ylianttila, M., & Liyanage, M. (2020). Survey on blockchain based smart contracts: Applications, opportunities and challenges. *Journal of Network and Computer Applications, 177,* 102857.

11. Crosby, M., Pattanayak, P., Verma, S., & Kalyanaraman, V. (2016). Blockchain technology: Beyond bitcoin. *Applied Innovation, 2*(6–10), 71.

12. Wood, G. (2014). Ethereum: A secure decentralised generalised transaction ledger. *Ethereum Project Yellow Paper, 151*(2014), 1–32.

13. Suma, V. (2019). Security and privacy mechanism using blockchain. *Journal of Ubiquitous Computing and Communication Technologies (UCCT), 1*(01), 45–54.

14. Dunham, I. (2016). Music information: the need for a central music licensing database. *IConference 2016 Proceedings,* University of Illinois, Champaign, IL.

15. Kishigami, J., Fujimura, S., Watanabe, H., Nakadaira, A., & Akutsu, A. (2015, August). The blockchain-based digital content distribution system. In *2015 IEEE fifth international conference on big data and cloud computing* (pp. 187–190). IEEE, Bulgaria.

16. Corrado, S. M. (2018). Care for a Sample: De Minimis, Fair Use, Blockchain, and an Approach to an Affordable Music Sampling System for Independent Artists. *The Fordham Intellectual Property Media and Entertainment Law Journal, 29,* 181.

17. Wikstrom, P. (2014). The music industry in an age of digital distribution. In Vazquez, J., Morozov, E., Castells, M., & Gelemter, D. (Eds.) *Change: 19 key essays on how the internet is changing our lives.* Turner/BBVA Group, Spain, 1–24.

18. Bao-Kun, Z., Lie-Huang, Z., Shen, M., Gao, F., Zhang, C., Yan-Dong, L., & Yang, J. (2018). Scalable and privacy-preserving data sharing based on blockchain. *Journal of Computer Science and Technology, 33*(3), 557–567.

# 2

## An Authentic Data-Centric Application for Medical Stores

**S. L. Jayalakshmi, R. Girija, and R. Vedhapriyavadhana**
*Vellore Institute of Technology, Chennai, India*

## CONTENTS

## 2.1 Introduction

Pharmacies use certain vital utilities to achieve their goals successfully. Technological support has now become obligatory to help pharmacists plan their stock. We often go to medical stores and pharmacies to buy medicine. Inside a medical store, the medical store inventory is managed by the staff and the proprietor [1]. Due to the outsized number of people inward-bound at these medical facilities, inventory administration and serving each customer becomes tedious tasks. Above all of that, the inventory management software or the bill generating software on these stores is usually a desktop application running on legacy and outdated technology. Thus, to overcome these challenges, a computational intelligence (CI) system is developed which will provide an up-to-date and efficient solution [2]. This eases a user-friendly environment and also an efficiently developed software application for medicines inventory management, and e-bill generation at medical stores and pharmacies to help the medical store staff to manage their store and customers in the best way [3–6].

Our app will keep track of stocks of all medicines, expiry dates of medicines, and other details such as the cost of a precise medicine. This technologically advanced software expedites functionality for locating a medication, especially in which particular section of the shop a medicine is kept. Additionally, it also affords the functionality of aiding the purchaser using the application by generating an e-bill and directly sending this to them. It will also trigger a reminder for the medicines which are going to be expired or out of stock. This app will ease the work of all medical store staff as well as the owner and make them process customer queries faster and handle them efficiently [4]. The intention of progressing toward such intelligent software is to offer a better medicine inventory management application and upgrade upon the flaws of the legacy software that are on the decline in the industry.

### 2.1.1 Purpose

The purpose of this application is to develop a user-friendly and efficient Android application for medicines inventory management at medical stores and pharmacies. This will solve the common problem faced by individuals especially at untimely hours is the unavailability of appropriate medicines. To address the same, a registered customer will prioritize a certain medicine based on the need. After successfully ordering it for the first time, the nearest pharmacy will receive a notification about the same. In addition to this, all the registered pharmacies will have a system of sensors for monitoring the inventory. Once stocks go below a certain level, the pharmaceutical company will be notified on an urgent basis for the same.

### 2.1.2 Intended Audience

The intended audience of the intelligent software applications will be hospitals, medical stores, pharmacies, medical store owners, and staff.

### 2.1.3 Product Scope

The scope of this application is different for small or big medical stores and pharmacy stores as well as hospitals that have to handle hundreds of medicines. The task of locating a particular medicine among many others can be tedious and time-consuming, which often results in chaos inside a medical store. This application will provide a handy app for medical store staff to locate a particular medicine by just entering its code or name and generating an e-bill that can be directly sent to the customer. In the meantime, it will also keep a check on the stock of the medicines and the expiry date of the medicine to lower the burden of the staff and owner.

### 2.1.4 Tech Stack Description

To build a fully functional intelligent application with a backend database and the ability to generate and email the e-bill of the customers. The Android app will also have a simple and elegant-looking user interface. The main Android application will be written using Java, XML, and Android Studio. The backend database will be built on Firebase.

### 2.1.5 Existing System

All the existing applications that we came along with while researching and developing this wonderful application were desktop applications that used outdated technologies and

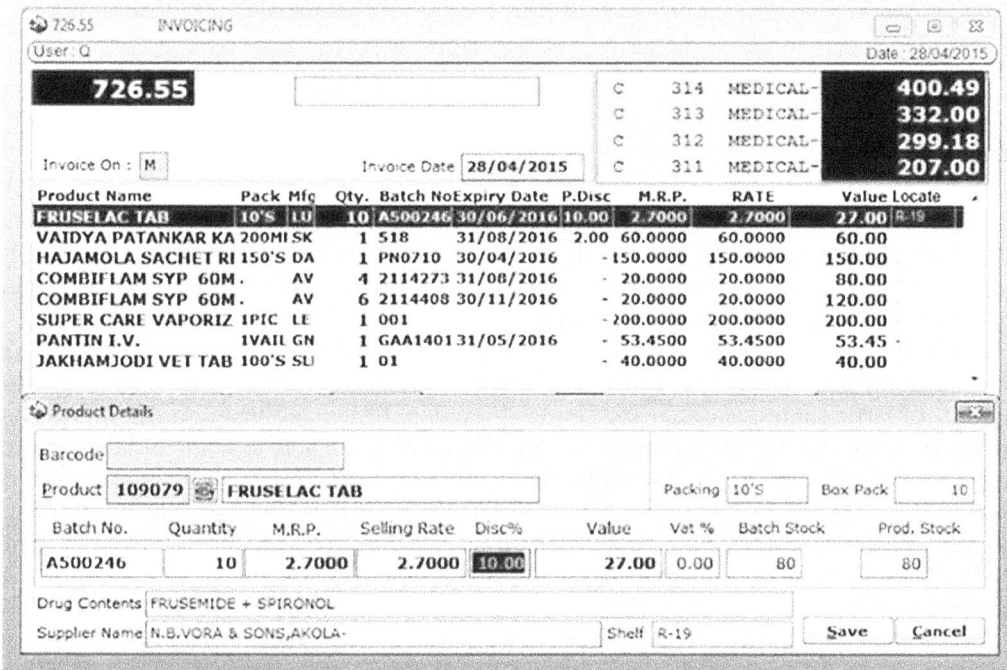

**FIGURE 2.1**
Conventional UI design desktop application.

had a very poor user interface. Below Figure 2.1 shows the existing system, which has a bad user interface (UI) design that limits the efficient use of the system. Limited ergonomic design restricts functional usage. As shown in Figure 2.1, UI below is too cluttered which can confuse the user and make them prone to error. The prevailing applications in existence are implemented as a desktop application as shown in Figure 2.1 which heavily limits their usage.

Limited accessibility reduces timely usage. Limited readability reduces timely interpretation. The problem with the desktop applications is that they are less portable, and each pharmacy or medical store can accommodate only one or two desktops. Desktop application also requires computer handling skills which most medical store workers lack. On the other hand, Android phones are more portable and almost everyone today has a smartphone running Android which means staff workers can easily use the Android application without any special skill requirement.

Existing systems do not provide any inbuilt payment method functionality; they just keep records and generate bills. Modern technologies have made digital payments and e-bill very popular and easy to implement and perform. The existing system built on legacy technologies are lack these features. These features if introduced can make the work of medical store staff a lot easier and faster. Keeping in mind all these drawbacks of the existing systems we decided to develop this Android application for pharmacy management which can perform inventory management (keeping records of medicine stock, expiry date, their shelf storage number, details, etc.) and at the same time provide the functionality for easy online payments and e-bill generation, which can be directly sent to the customer's email. This will help to reduce the workload at medical stores/pharmacies and provide them with a better and more efficient solution for their store management using modern and up-to-date technologies.

## 2.2  Literature Survey

The Hospital Management System is commonly used to enhance the operation of hospitals in the medical field. However, there is a privation of a function to provide a prescription information service for clinical pharmacists.

In [1], the clinical pharmacy management system (CPMS) software has been used to provide drug information systems, encourage abnormal medicine use, and carry out surveys of large sample cases using the urology department as an example. CPMS has been used to test medications (medical orders) that have been able to warn and respond to unreasonable prescriptions in a timely fashion, to have direct contact with physicians, and to help discourage irrational drug use. CPMS was related to the hospital information system (HIS), which was able to openly access medical information of patients and their medication (medical order) information through online mode, thereby decreasing the amount of manual time used up by the clinical pharmacists in the department.

The author of [2] reviewed the various pharmaceutical operations in the inventory control systems. Their research mainly focused on the stock management of the hospital to maintain the demands of the medical products. Efficient resource management plays a crucial role in pharmaceutical treatment from a financial and organizational point of view. Inventory control aims to reduce sourcing and billing costs while at the same time maintaining a productive supply of goods that meets customer and prescriber needs. The author outlined various approaches and roles of information technology with various factors for inventory negligence on treatment and rehabilitation.

The work in [3] discussed the case study of UNTH Enugu university hospital to address the need for creating a method that will enhance the categorization of stored pharmaceutical products in the pharmacy. Additionally, this method reduced the certain losses faced by the existing operations of the pharmacy software. Drug recognition and delivery monitoring software contains a collection of programs to gather stocks of drugs, sell drugs, and manage the regulation of drug inventories. Previously, the department's manual mode of service lacked in storing and maintaining the medication information system. In addition to this, the recording system for tracking expired drugs is also not good. A successful solution for the above-mentioned problems was developed with a help of C# programming language to do the following operations: (i) creating a form; (ii) providing access to the database; (iii) maintaining a correct record of drugs, and (iv) to avoid the sale of expired drugs. It permits people for using better health services and assures that the medications they consume are genuine and secure. Drug production, tracking, supply, and data processing are routine procedures in different hospitals across the country. Thus, the proposed computer-based system has assisted to manage the drug information and distribution tracking of pharmaceutical products.

The authors in [4] demonstrated the Pharmacy Management System with a stock alert system. This system improved the performance of existing software by providing secured access for the drug products. After that, the system provided a sophisticated environment to execute some sort of tasks in a procedural way for controlling much of the pharmaceutical operations in the pharmacy. This system does not have a mobile application that is easily accessible to the general public than a desktop app, and the UI offered is not aesthetically appealing either. Nor does this app have a built-in payment system.

In [5], the author provided a variety of insights into the nature and implementation of the Pharmacy Management System. This system offers some sort of activities in a specialized manner. For example, it is used to control a significant portion of the pharmacy's operations. This software assists pharmacists in improving inventory control, pricing, patient safety, and so on. Between the opening of the stock and during the sales [6–9], the customer can get the date of manufacture and expiry for a certain product or substance. The system will also provide a report that shows the list of items that expire after the specified period before the product's expiration date. It also necessitates manual entry on the arrival of new collections of drugs and the transport of drugs out of the pharmacy over a certain period (e.g., every month). Additionally, the pharmacist needs to be given updates on the movement of drugs into and out of the pharmacy and receive information about the drugs, e.g., expiry date, date purchased, the amount of drug type left, location of a drug in the pharmacy. This software replaced the manual method used in pharmacies by encouraging pharmacists to automate the monitoring of all medications on the premises [10]. The biggest drawback of this scheme is that the creator did not make the UI esthetically pleasing, and there is also no payment portal offered to customers [11–14].

In [15], proposed work challenges to expansion a recovering empathetic on the central subjects distressing the administration of tablets in infirmaries by essential and examining the foremost logistics ineptitudes effected by the hospital drugstore. This field exploration was shown in a North American infirmary which epitomizes the foremost examination place. Three other clinics and other healthcare governments have been complicated in command to authenticate experimental suggestions. It has been trusted on manifold foundations such as explanations, procedure charting, and semi-structured meetings in command to agree on triangulation and fortify the legitimacy of consequences. As of the response of medications at infirmary ports to their spreading to the clinic drugstore, it has been recognized as a significant total of ineptitudes, of explicitly improper catalog controlling, remedy reduction, concentrated physical industry, extended obtaining sequences, time-consuming invention with indecorous usage of expertise.

In [16], pharmacy inventory organization is a life-threatening procedure in healthcare centers. On the one hand, real drug gaining is important for rewarding the relaxing rations of patients. On the supplementary arrow, as hospital drugstores' procuring also packing outlays embrace a significant part in the infirmary finances, well-organized catalog controlling can production a significant character in operating cost restraint. So, healthcare middle people must project and use tools for decision-making of medication acquisition, with the intention of dodging extreme bulk-buying and raising storeroom volume, though also anticipating patient demand and safeguarding against dangerous storage levels. This study focuses on the current procedural topographies of a decision-aid instrument for the development of acquisitions and list levels for the medicine drugstore of the Local Hospital of Talca, Chile. The report and the consequences gained after 1 year of process; these consequences demonstration that our plan formed more than 7% reserves associated to the steady record organization scheme and was more operative in protective critical routine intensities. Furthermore, from a computational opinion of interpretation, our approach outclasses a newly printed method for an alike request.

To solve all the problems in the state-of-the-art systems [17–20], we proposed a system to use the technique for supervising the pharmacy using a CI approach.

## 2.3 Proposed Methodology for a Data-Centric Approach in Medical Stores

Figure 2.2 shows the architecure diagram of data centric approach in medical store in software modeling. This architecture clearly illustrates the ten substances which are needed for this supervision. This architectural design consists of three roles: Administrator, Customer, and Staff. The main roles of the customer are ordering medicine, viewing bills, and making payments. The responsibilities of the administrator are to manage staff, manage stock, and manage the database. Finally, the obligations of the staff are finding medicines, adding medicines, viewing stock, and generating bills and receipts.

   i. Consumer Sector: Customers are the key component of this application. Some tasks in the customer module are ordering medicines, viewing bills, and making payments.
   ii. Artifact Sector: When a complete pharmacy system is managed with the use of CI-based upon software, implicitly items are updated such as addition, deletion, replication, neglecting, etc.
   iii. Item Kit Sector: If the item module is verified and validated in an enormous mandate, it can be done using the item kit module.
   iv. Distributor Sector: Supplier acts as a bridge between customer and staff.
   v. Reporter Sector: In software modeling, the reporter module is a crucial module. This module collects and stores all the details such as customer satisfaction, supplier satisfaction, verified and validated items.

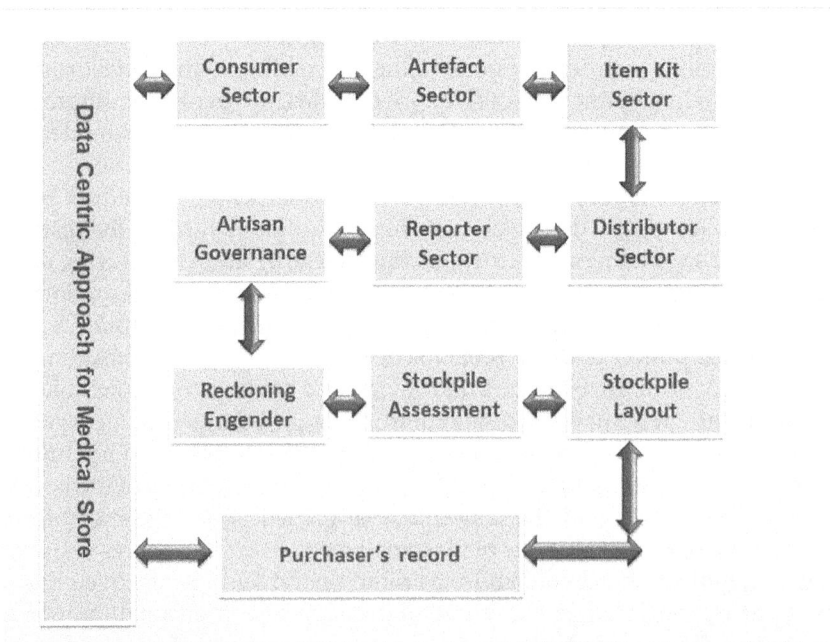

**FIGURE 2.2**
Architecture diagram of data-centric approach for medical store.

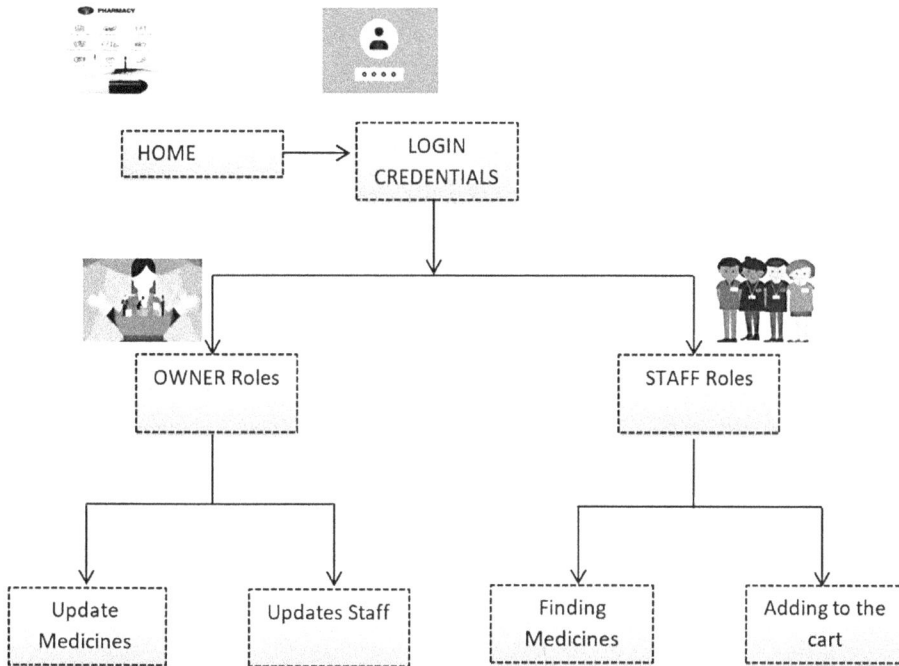

**FIGURE 2.3**
Roles and responsibilities of owner and staff in pharmacy supervision based on computational intelligence.

vi. Artisan Governance: Since this makes the best use of CI, employee management is automatically updated.

vii. Reckoning Engender: Pharmacy management system is configured properly and is well-stored.

viii. Stockpile Assessment: CI for pharmacy-based application design automatically checks and evaluates the stock.

ix. Stockpile Layout: This application verifies everything and finally generates the invoice for the customer.

x. Purchaser's Record: Moreover, this pharmacy application based upon CI creates, generates, verifies, and updates the customer record.

The above Figure 2.3 depicts the navigation page from home screen, checking the credentials and responsibilities of pharmacy application in computational intelligence in software modeling.

### 2.3.1 Implementation of Intelligent Pharmacy Application

The initial step for this intelligence based pharmacy application is to create a firebase account. Figure 2.4 shows the construction of firebase account.

The secondary step is to add the project button in order to create and register the app in Firebase. After the second step, enable all the Google analytics services in the third step. Figures 2.5 and 2.6 show the second step and third step, respectively.

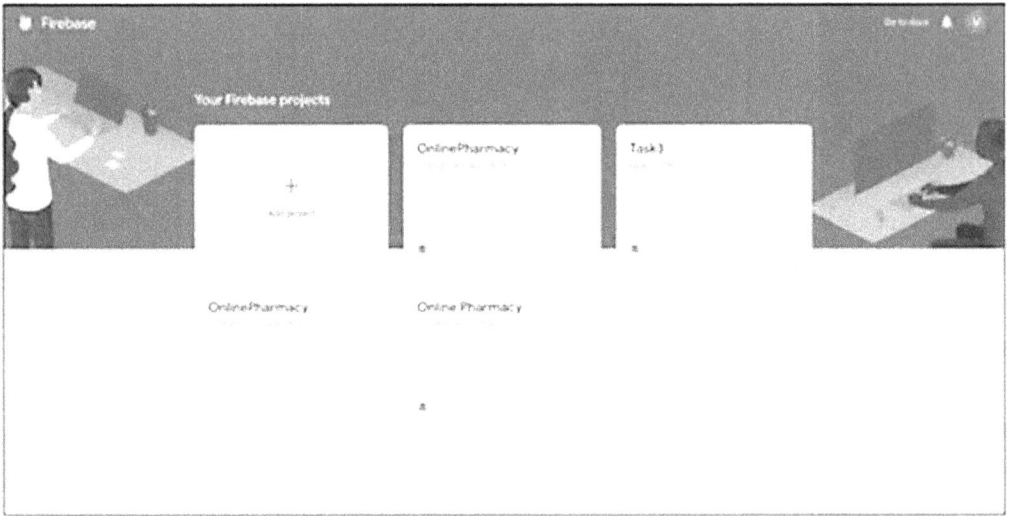

**FIGURE 2.4**
Account creations in firebase.

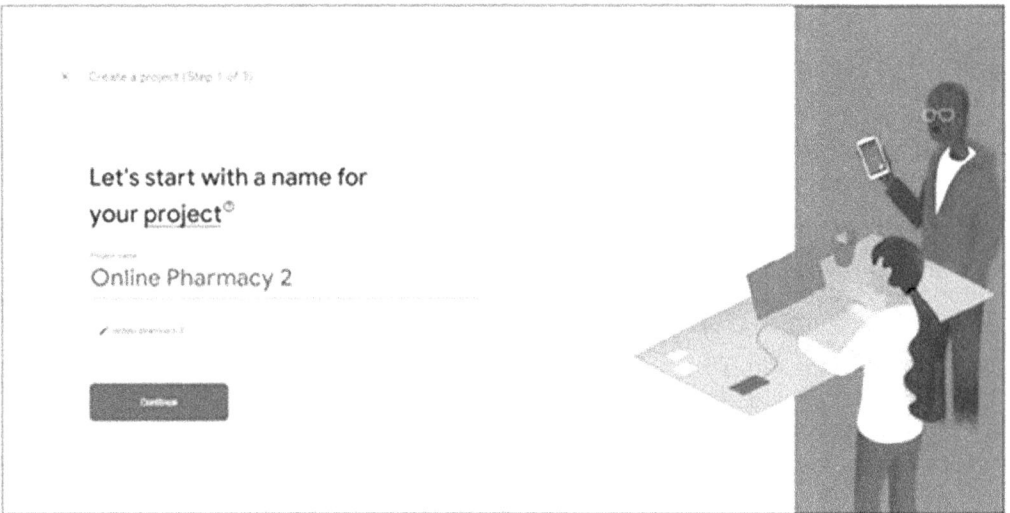

**FIGURE 2.5**
Adding project in Firebase.

The fourth step is to choose your location and accept all the terms and conditions needed by Firebase, then click *Create project*. It's shown in Figure 2.7 and done by configuring Google analytics. Finally, the project has been created and is represented in Figure 2.8.

After creating the project, it is mandatory to register the Android app to the Firebase. Copy the package of your Android project and paste it in *Register app field*. By adding the name of the project, then add the register app; then download the google-services.json file and add it to the project, which is mentioned in Figure 2.9.

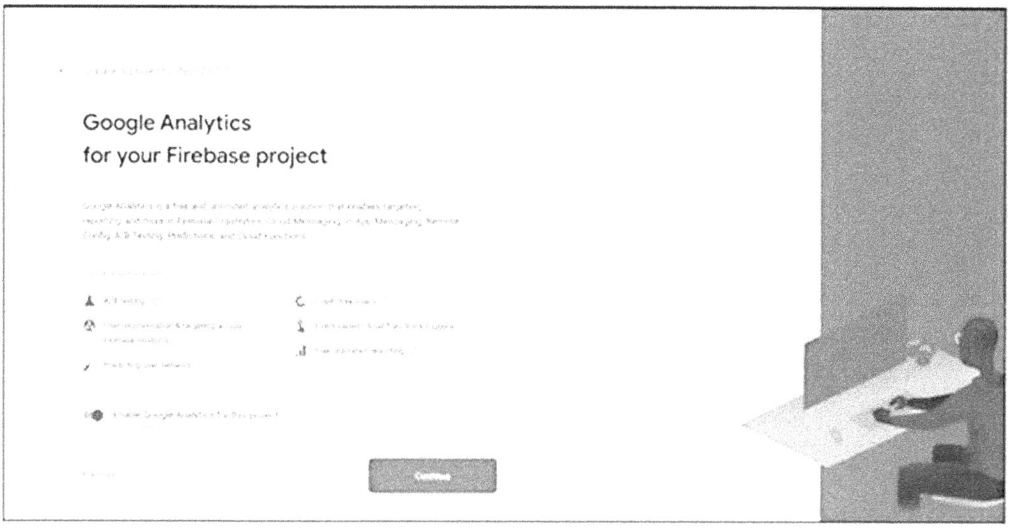

**FIGURE 2.6**
Enable Google analytics.

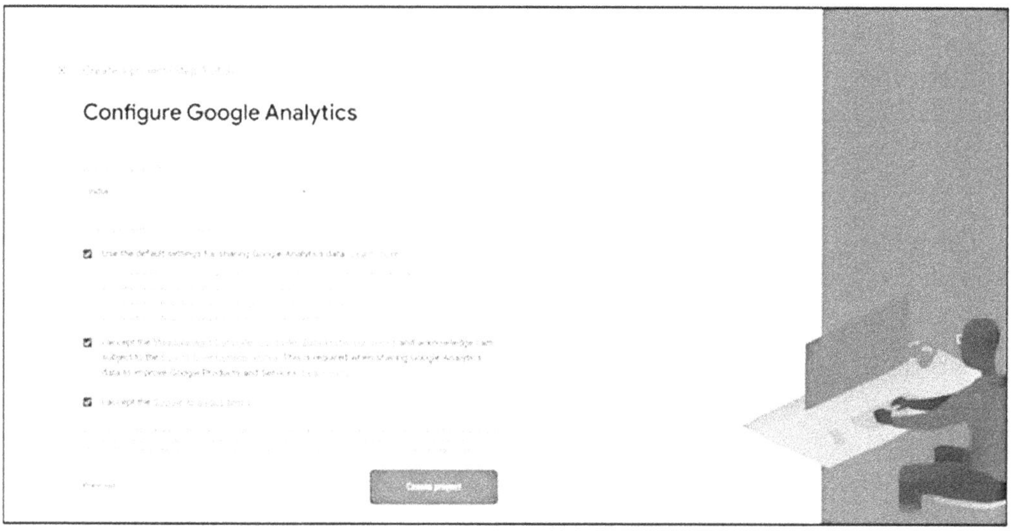

**FIGURE 2.7**
Configure Google analytics.

In order to check the authentication of the user, Firebase authentication is declared which securely saves user data in the cloud and provides backend services, an easy-to-use software development kit (SDKs), and a ready-made UI library to authenticate users to your app. Figure 2.10 represents the user authentication in Firebase.

To synchronize each updated data in real-time for every connected client, a Firebase Realtime Database has been created (see Figure 2.11). In order to store the data, it is necessary to make class to mimic the structure of the data and its attributes and then make the

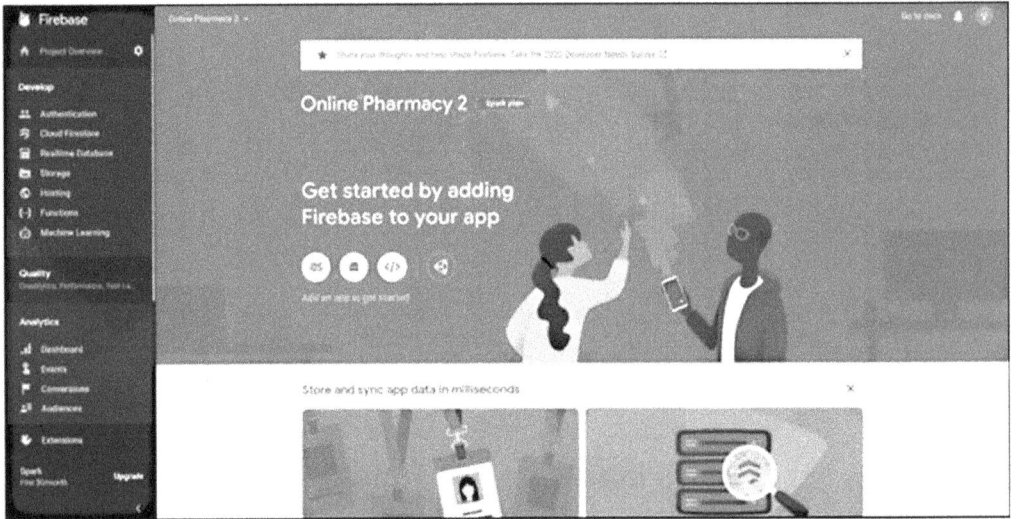

**FIGURE 2.8**
New project creations.

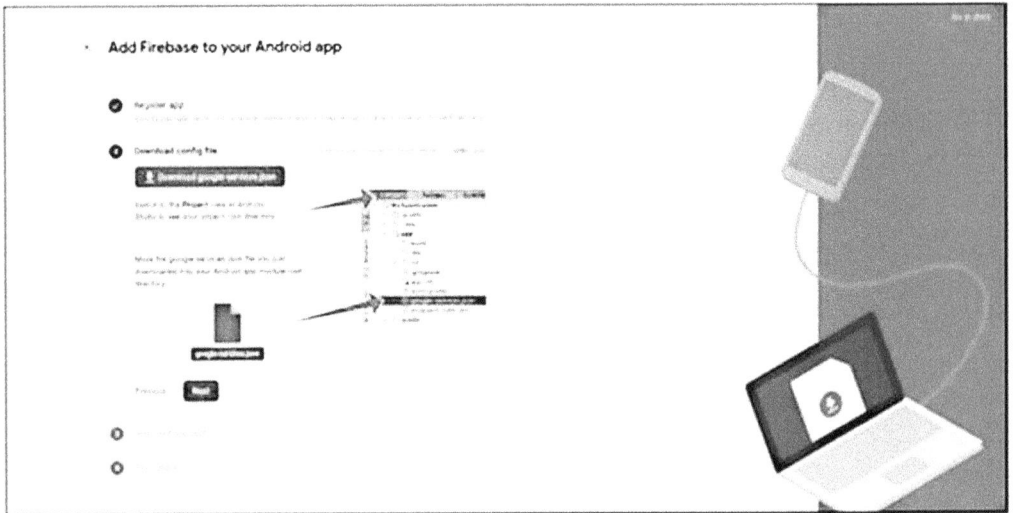

**FIGURE 2.9**
Downloading the config file.

object of that class push it into the Firebase Realtime Database where the whole object is converted into a JSON tree and stored. Data storage and conversion are shown in Figure 2.12.

After performing a deeper analysis in implementation and its results, the pharmacy supervision application which is designed on the ideology of CI in software modeling is proved to be a better design.

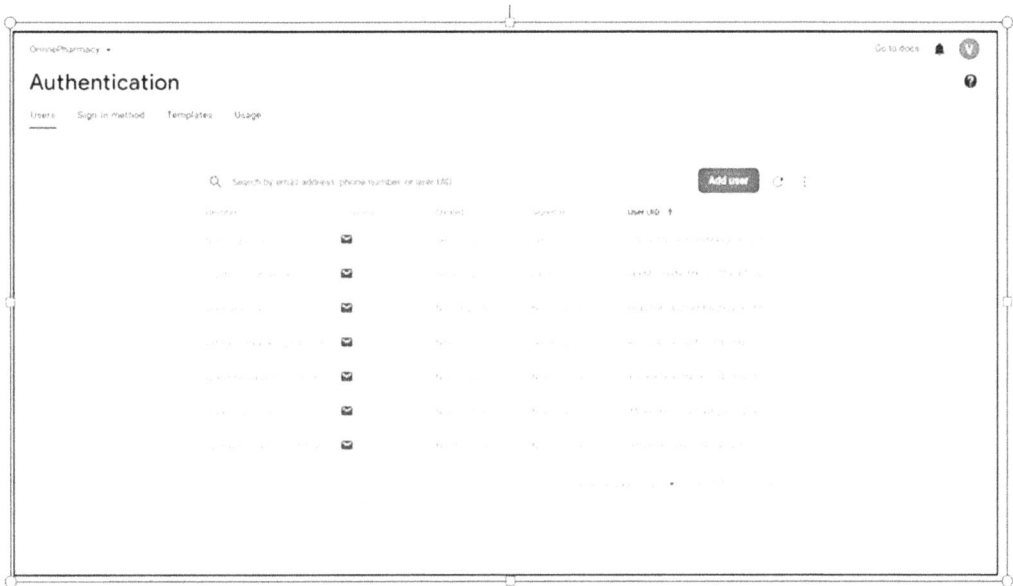

**FIGURE 2.10**
User authentication.

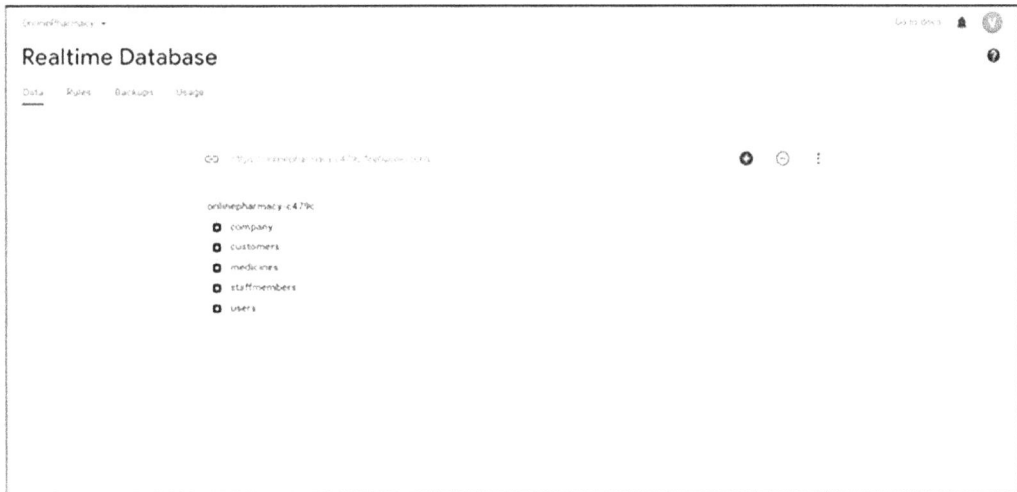

**FIGURE 2.11**
Creation of Firebase real-time database.

## 2.3.2 Results and Discussion

In comparison to the previous system [1,2,10], this application is very useful for the inventory system, serving consumers, and creating and delivering e-invoices directly to them by medical stores, pharmacies, and hospitals. This system provides reliable control on the medication system and prevents medicine deception.

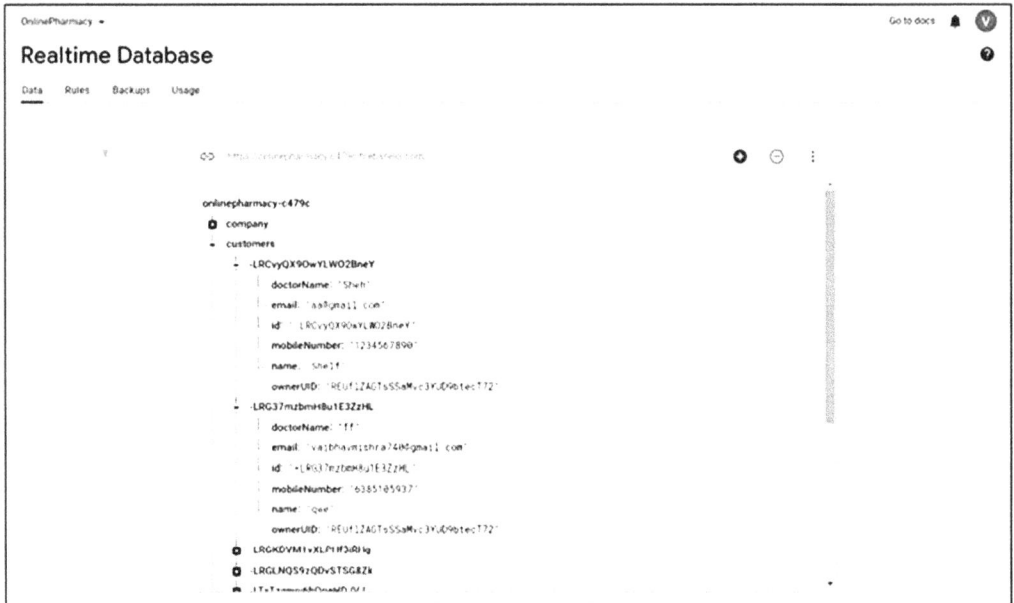

**FIGURE 2.12**
Storage and its conversion of Firebase real-time database.

## 2.4 Conclusion

As we have learned in the above-described application, the implementation of an Android application for pharmacy management can be done effectively. This application is built with modern technologies such as Android (Java and XML) and Firebase (Database). The UI of our application is simple, highly esthetic, and efficient in use. This Android application performed its entire task as per the requirements. This application can be used for inventory management, as well as for serving the customers and generating and sending e-bills directly to them by the medical stores, pharmacies, and hospitals.

## References

1. Ren, X., Wang, Y., & Ma, R. (2016). A novel clinical pharmacy management system in improving the rational drug use. *Clinical Medicine Research*, 4(6), 229.
2. Ali, A. K. (2011). Inventory management in pharmacy practice: A review of literature. *Archives of Pharmacy practice*, 2(4), 151.
3. Khanna, D. (2018). Use of artificial intelligence in healthcare and medicine. *International Journal of Innovations in Engineering Research and Technology*, 5(12).
4. Sarker, N. C. (2018). Online Inventory and Order Management System for Pharmacy (M. Sc Thesis)
5. Zangana, H. M. (2018). Design an information management system for a pharmacy. *System*, 7(10).

6. Brown, S. L., De Jager, D., Wood, R., & Rivett, U. (2006). A pharmacy stock control management system to effectively monitor and manage patients on ART. 27–35.

7. Hong, T., Dong, M., Zhao, J., Fu, X., & Chen, Y. (2012, August). The application of information technology in the hospital pharmacy management based on HIS. In *2012 International Symposium on Information Technologies in Medicine and Education* (Vol. 2, pp. 604–607). IEEE.

8. Hashim, A. S., Yusoff, M. F. M., Sarlan, A., Mahamad, S., & Basri, S. (2016, August). Development of myhomepharmacy: A personalized family medicine management. In *2016 3rd International Conference on Computer and Information Sciences (ICCOINS)* (pp. 611–615). IEEE.

9. Williams, F. L., & Nold, E. G. (1983, October). Selecting a pharmacy information system. In *The Seventh Annual Symposium on Computer Applications in Medical Care, 1983. Proceedings*. (pp. 200–202). IEEE.

10. Bao, L., Wang, Y., Shang, T., Ren, X., & Ma, R. (2013). A novel clinical pharmacy management system in improving the rational drug use in department of general surgery. *Indian Journal of Pharmaceutical Sciences*, 75(1), 11.

11. Basha, M. J., Navya V. S., Tukaram Wani, S., & Vivekanand, S. G. (2020). Study of inventory management in pharmaceuticals: A review of COVID-19 situation. *International Journal of Innovative Science and Research Technology*, 5(8), 366–371.

12. Bialas, C., Revanoglou, A., & Manthou, V. (2020). Improving hospital pharmacy inventory management using data segmentation. *American Journal of Health-System Pharmacy*, 77(5), 371–377.

13. Khelifi, A., Ahmed, D., Salem, R., & Ali, N. (2013). Hospital-pharmacy management system: A UAE case study. *International Journal of Computer, Control, Quantum and Information Engineering*, 7(11), 795–803.

14. Huang, S., Axsäter, S., Dou, Y., & Chen, J. (2011). A real-time decision rule for an inventory system with committed service time and emergency orders. *European Journal of Operational Research*, 215(1), 70–79.

15. Bush, J. K. (1984) A method of assessment for hospital pharmacy stock control computer systems. In: van Eimeren W., Engelbrecht, R., Flagle, C.D. (eds) *Third International Conference on System Science in Health Care. Health Systems Research*. Springer, Berlin, Heidelberg. doi:10.1007/978-3-642-69939-9_170.

16. Holm, M. R., Rudis, M. I., & Wilson, J. W. (2015). Medication supply chain management through implementation of a hospital pharmacy computerized inventory program in Haiti. *Global Health Action*, 8(1), 26546.

17. Woo-Miles, K. (2015). Evaluating hospital pharmacy inventory management and revenue cycle processes: White paper guidance for healthcare internal auditors. *Costa Mesa, Deloitte*. https://ahia.org/assets/Uploads/pdfUpload/WhitePapers/EvaluatingHospitalPharmacyInventoryManagementandRevenueCycleProcesses.pdf

18. Romero, A. (2013, October). Managing medicines in the hospital pharmacy: Logistics inefficiencies. In *Proceedings of the World Congress on Engineering and Computer Science*, San Francisco, USA (Vol. 2, pp. 1–6).

19. Silva-Aravena, F., Ceballos-Fuentealba, I., & Álvarez-Miranda, E. (2020). Inventory management at a Chilean hospital pharmacy: Case study of a dynamic decision-aid tool. *Mathematics*, 8(11), 1962.

20. Fernández, M. I., Chanfreut, P., Jurado, I., & Maestre, J. M. (2020). A data-based model predictive decision support system for inventory management in hospitals. *IEEE Journal of Biomedical and Health Informatics*, 25(6), 2227–2236.

# 3

# Intelligent Data-Analytic Approach for Restaurant Recommendation

**Nirmal Kirola, Sandesh Tripathi, and Nitin Chhimwal**
*Birla Institute of Applied Sciences, Bhimtal, Uttarakhand, India*

**Sudhir Sharma**
*Manipal University Jaipur, Rajasthan, India*

## CONTENTS

## 3.1 Introduction to Machine Learning

As we see, it is really hard to travel anywhere now and not hearing about topics like computing and machine learning and data. Data is especially everywhere [1]. Research has suggested that every two years we generate more data than has ever existed before.

Therefore, we are able to say that the number of data is doubling every two years. Now, that is an astronomical amount of information.

However, there is a thing to notice that this data does not necessarily mean anything that you simply can create tables of knowledge, but unless you understand what is in them and what they mean, you do not have any knowledge. As a result, there is a distinction to be made between data and knowledge [2]. We generate a lot of data, but we do not use a lot of it. A lot of stuff is resting on a hard disk, waiting for someone to look at it, and that is probably what we want to discuss though. If we wish to extract knowledge from data, we are going to need some tools and processes to try to do this in a formal way, that is what they decided to try to do, and things like machine learning and artificial intelligence (AI) have a pace within it [3]. To say we are taking plenty of different data sources and dealing out the optimal route. This type of stuff you are already doing it. It is just a case of trying to formalize this process.

Well, one problem is that everyone's definitions differ slightly, but also I believe that a lot of those terms are used completely interchangeably. AI is a classic example. AI is everywhere, right? We say you cannot buy a product now without it having inbuilt AI. Much of the time, you see we are actually talking about machine learning [4]. Machine learning is the idea of teaching a machine to complete a task without explicitly programming it to do so. An honest example of what is machine learning would be a mouse in a maze, where all you are doing is directing it to go left or right at random, without learning anything. It has no idea what the maze is, but it will ultimately find its way out. This is fundamental computing, but it does not require knowing anything. Machine learning is about refusing to give up. The study of formal language induction, often known as grammatical inference, is a hot topic in learning algorithms and computational learning theory [5,6].

Machine learning may be a subset of AI but it should not be used interchangeably. If we are using machine learning often, what we will do is we will train it with supported samples of knowledge [7]. Therefore, we will have some existing data set that we are trying to coach on and we are trying to use the machine learning to either tease out information or make predictions on this data. The matter is that not all data is variety of made equal a number of its noisy and messy. Maybe we do not know what it is and do not know whether we will apply a particular technique to that. We want to scrub this data for a much better analysis. We have taken this data understand what it is and extract some knowledge so we are able to then apply these AI or machine learning techniques to that. This mix of things that may take data preparing how that we will then use it or comprehend it that is data science.

We are going to analyze the info still going to do data analysis and maybe sometimes just using statistics to research the information is not enough, you cannot really learn everything about it. Yes, you will be able to learn, you know, mathematically how it works, but you would possibly not understand about what it all means. Visualizing the information will be helpful [8]. So, what we also do is visualize the info involved in plotting the trends and links between different variables. Things like this can be quite back and forth. You could do both of those things numerous times and see what we have. So you are going to try to something like this and we are going to do is we are going to pre-process the data often [9]. You will be finding your recording far more data than you truly need. This is often certainly true and is also a component of this really.

What data processing could be a combination of pre-processing your data and perhaps using clustering to extract some knowledge from it [10]. A word come to be utilized in place of these things. If someone says they are doing data processing, that is what they are doing [11]. They are pre-processing and extracting some knowledge from the information. You have collected many samples of something and now you are analyzing it [12].

Some reasons for using data science technology:

i. To transform the huge amount of raw and unstructured data into meaningful insights.
ii. Opted by various companies.
iii. Companies using data science algorithms for better customer experience.
iv. Functioning for automated transportation like self-driving car.
v. Helps in numerous predictions like elections, various survey, flight ticket confirmation, etc.

## 3.2 Related Work

### 3.2.1 Exploring Venue Popularity

This work is the detailed analysis in on the venue quality in Foursquare. It focuses on finding out the common characteristics of widespread venues. The main motive of this analysis is to grasp the factors that are the explanations for venues to become widespread. It certainly has implementations in suggestion of venues [13].

### 3.2.2 Software Fault Prediction

The work of software fault prediction is using Quad Tree-based K-means rule. The task is to predict faults in the program modules. The error rates calculated through this approach are compared to utterly completely different algorithms. Results are observed to be higher in the majority of cases [14].

### 3.2.3 Temperature and Humidity Data Analysis

According to this paper on temperature and humidity data analysis, knowledge of climate data in a region is important for business and other applications. Various data mining techniques such as prediction, clustering, classification, and outlier analysis are used. This work aims to fetch temperature and humidity data and clustering it for climatic condition classification [15].

### 3.2.4 Educational Data Mining for Prediction of Student Performance

It is observed for a few years that the rapid growth of educational data is a challenge. Using this data is also a challenge. As data contains parameters affecting students' performance, it becomes crucial to utilize this data in an efficient manner. Data mining techniques could be useful in classifying data. This paper aims to develop a trust model, which helps in achieving the above things [16].

### 3.2.5 Crime Prediction and Forecasting

Crimes are most prevalent in society and need to be prevented from happening. Crime analysis is a structured way of detecting trends and patterns in crimes. Clustering algorithms are compared to find the best suitable clustering algorithm for crime detection. After look-ing at the results, K-means clustering is used for clustering. The K Nearest Neighbors clas-sification is used for crime prediction [17].

### 3.2.6  Ship Trajectory Prediction

Ships are required to guide when they pass through controlled waterways. Inaccurately predicted trajectories may cause traffic jams and traffic jams on water could be horrible. The historical data of trajectories are bunched by the K-means algorithm, and then artificial neural network (ANN) models are built to predict the ships' trajectories. It will help largely to generate the optimal traffic commands [18].

### 3.2.7  Incident Analysis and Prediction

This study advances in the infrastructure of data collection offer various opportunities. A system could be built which can foresee and respond to incidents that are heteroge-neous in nature. There is incident forecasting, which performed using previously collected vehicular accident data is provided by Nashville Fire Department. SBAC analysis is used to categorize incidents. Survival analysis is used to determine the likelihood of incidents per cluster. The Bayesian network is used for the mapping of clusters. The model that is built lays the foundation for an optimal and much more accurate emergency vehicle alloca-tion and dispatch system [19].

### 3.2.8  Cross-Validation of Protein Structural Class Prediction

This paper defines an approach to predict protein structure class. A three-layer back-propagation network and learning vector quantization network is designed. Results are compared to those of the Euclidean statistical clustering algorithm. This tells us that infor-mation exists in primary sequences of protein, just need to utilize it [20].

## 3.3  Research Methodology – from Collecting Data to Suggesting Places

The below diagram shows all our processes which we will be completing in sequence (Figure 3.1).

### 3.3.1  Data Sources (Our Data Points)

In order to get the whereabouts and other details about various food venues in Mumbai, two APIs are used and decided to combine the data from both of them together.

Using Foursquare's API (which gives the suggestion for venues), venues are fetched to the range of 4 kilometers from Mumbai's center and aggregated their names, locations, and categories [21].

Using the name, longitude, and latitude values, the Zomato API is used to fetch venues from its database. This API helps to find venues based on search criteria (name), latitude and longitude values, and more. Data cleaning is used to combine the two datasets properly.

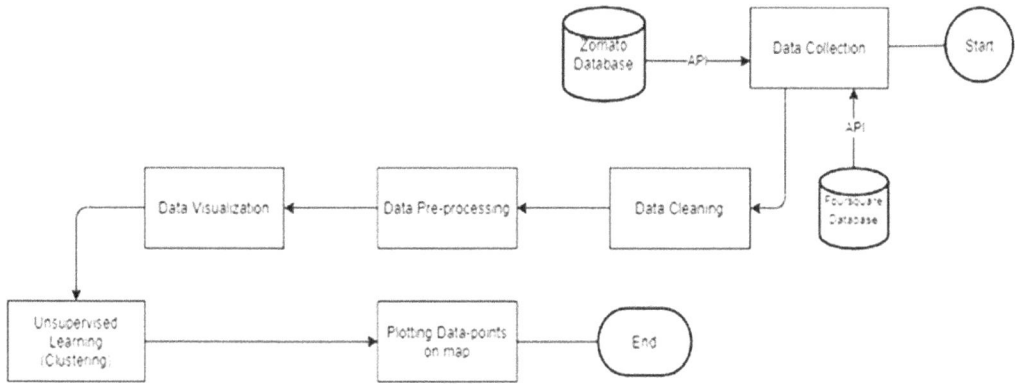

**FIGURE 3.1**
Dataflow process of the proposed model.

From the Foursquare API (https://developers.foursquare.com/api) [22], the following data is retrieved for each venue:

  i.   Name: The name of food venues.
  ii.  Category: The section type as characterized by the API.
  iii. Latitude: Latitude for the food venue.
  iv.  Longitude: Longitude value for the food venue.

On the other hand, From Zomato API (https://developers.zomato.com/api) [23], the following data is retrieved for each venue:

  i.   Name: Name of the food venue.
  ii.  Address: Complete address of the venue.
  iii. Rating: The ratings as provided by many users.
  iv.  Price range: The price range the venue belongs to as defined by Zomato.
  v.   Price for two: Average price per person.
  vi.  Latitude: Latitude for the food venue.
  vii. Longitude: Longitude for the food venue.

### 3.3.2 Analysis of the Venues Retrieved (Formatting and Cleaning)

From Figures 3.2 and 3.3, we can clearly see that some venues from the two APIs do not align with each other. Thus, these venues are combined using their latitude and longitude values. To combine the two datasets, we have to check that the latitude and longitude values of each corresponding venue match. After careful analysis, we decide to drop all corresponding venues from the two datasets that have their latitude and longitude values differ by more than 0.0004 from one another. Thus, both the latitude and longitude values are rounded up to four decimal places. Then, the difference between the corresponding latitude and longitude values is calculated and see if the difference is less than 0.0004, which should ideally mean that the two locations are the same. This will remove many outliers

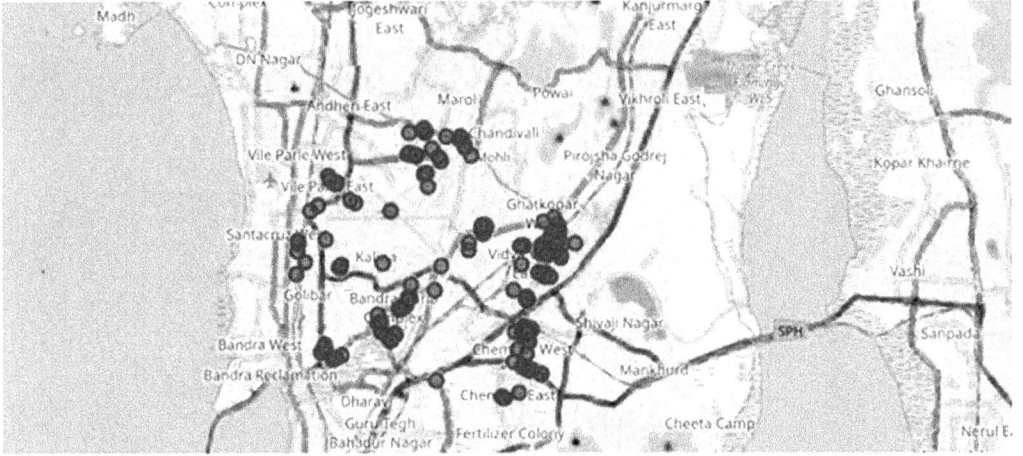

**FIGURE 3.2**
Venues retrieved from Foursquare API.

**FIGURE 3.3**
Venues retrieved from Zomato API.

from the two datasets. Once this is done, it is observed that there are still some venues, which are not correctly aligned. They can be categorized as follows: there are venues that have specific restaurants/cafes inside them as provided by Zomato API and two locations are so close that they have practically the same latitude and longitude values.

### 3.3.3 Some Venues Have Been Replaced with New Venues

Venues belonging to categories 1 and 3 are perfect to keep. However, the venues that belong to category 2 should be dropped. After careful inspection and removal, the final dataset would have 49 venues with which we can work. As a final dataset, we are left with 75 venues with eight columns as described in Figure 3.4.

| | categories | venue | latitude | longitude | price_range | rating | address | average_price |
|---|---|---|---|---|---|---|---|---|
| 0 | Hotel | Tuskers - Sofitel | 19.0673 | 72.8692 | 4.0 | 4.0 | Sofitel Hotel, C 57, Bandra Kurla Complex, Mumbai | 1650.0 |
| 1 | Hotel | O22 - Trident | 19.0672 | 72.8675 | 4.0 | 4.0 | Trident Hotel, C 56, G Block, Bandra Kurla Com | 2000.0 |
| 2 | Indian Restaurant | Masala Library | 19.0690 | 72.8696 | 4.0 | 4.4 | Ground Floor, First International Financial Ce | 2500.0 |
| 3 | Ice Cream Shop | Natural Ice Cream | 19.0776 | 72.8628 | 1.0 | 4.1 | 3, Gokul Harmony, Kalina Market, Sunder Nagar, | 150.0 |
| 4 | Deli / Bodega | Smoke House Deli | 19.0688 | 72.8695 | 4.0 | 4.3 | 3A, Ground Floor, 1st International Financial | 1250.0 |

**FIGURE 3.4**
Final data aggregated from both APIs.

### 3.3.4 Clustering – Finding Similar Places

Clustering is a method, technique, or approach, which is used for data analysis using which we can get a comprehension about what the composition of the data looks like [24]. We can think of it as of grouping the data such that data points, which share some common attributes, are grouped together. Each cluster is different from every other cluster. There are two types of similarity measures, one is Euclidean-based distance and the other one is correlation-based distance [25]. The decision to choose similarity measure completely depends on the purpose of application [26]. There are many applications where clustering can be used, let us take one example as market segmentation, where customers are divided into clusters on the basis of their behavior or attributes [27]. Another example we can say is in image segmentation/compression and in document clustering.

Clustering is what we can say, is a type of an unsupervised learning because here we are not comparing the output, which we get through the clustering algorithm to the labels for assessing the performance of the algorithm. We only focus on the composition of the data by bunching the data points into distinct clusters or subgroups [28].

We will now study about K-Means clustering, which is one of the widely used clustering algorithm due to its simple nature and is perfect for our purpose.

### 3.3.5 K-Means Clustering – the Hot Approach

K-means clustering is a non-parametric method of clustering, non-parametric means computational complexity of the algorithm depends completely on the number of samples. The K here is how many clusters we ever want. This is defined by setting K in K-means. It is a tunable hyper-parameter which is a parameter, that is set by the programmer manually and the algorithm does not learn it. In summary, K-means is a non-parametric method of clustering where we predefine the number of clusters [29].

Let us move to functioning of algorithm:

1. Determine the no. of clusters K in which we want to determine the data.
2. Initialize K random non-overlapping points as the cluster centers. We can select these points from our data points.
3. Assign each of the N data points to the closest cluster center.
4. For every cluster, compute the centroid.
5. Centroid is the new cluster center.
6. In final step, we just go back to step 3 where we repeat it until convergence.

After successive iterations of the algorithm the cluster centroids do not change that much. Then we know that the algorithm has converged. We will end up N data points bunched into K clusters.

Now one question arises when are computing the cluster center; Why exactly are we considering the centroid? Well, in order to understand that will have to go through the math and let us do that right now:

As with any type of statistical learning method, the idea is to optimize some objective or minimize a loss for K-means. This loss is the Distortion function, as shown in equation 3.1.

$$J = \sum_{i=1}^{m} \sum_{k=1}^{K} w_{ik} \left\| x^i - \mu_k \right\|^2 \tag{3.1}$$

In plain English, this Distortion function is the sum of distances of every sample from its cluster center. In this, the form $J$ is the Distortion measure as shown in equation 3.2, $x^i$ is the $i^{th}$ sample, $u_k$ is the $k$th cluster center and $w$ represents the cluster membership here. It is an $i$ cross $k$ matrix with values either 0 or 1. $W_{ik}$ is equal to 1 for only that value of $k$ where the $k$th cluster center is the closest. Otherwise, it's zero as shown in equation 3.3. Therefore, you can see that every row is actually a one hot encoded vector. In $k$-means clustering the only set of parameters we need to learn are the cluster centers. The idea is to find the optimal cluster centers UK that minimize the Distortion.

$$\frac{\partial J}{\partial w_{ik}} = \sum_{i=1}^{m} \sum_{k=1}^{K} w_{ik} \left\| x^i - \mu_k \right\|^2 \tag{3.2}$$

$$\Rightarrow w_{ik} = \left\{ 1 \quad if \quad k = argmin_j \left\| x^i - \mu_k \right\|^2 ; \quad 0 \quad otherwise \right\} \tag{3.3}$$

We determine this by taking the derivative of Distortion with respect to $u_k$ and equating it to 0 as shown in equation 3.4. Let us split the sigma over the terms. Now remove the common too. Take $u_k$ out of the sigma as it is independent of $i$ and brings all the terms to the right-hand side. In English, what does this represent? Well, the numerator is the sum of all samples $xi$ belonging to the cave cluster while the denominator is just the number of samples in the cluster $k$ as shown in equation 3.5. This sum of all samples in a cluster divided by the number of samples in the cluster is the definition of centroid. The reason why we are determining centroids in order to define the cluster centers is now cleared. Further, it actually gives us meaningful results now [30].

$$\frac{\partial J}{\partial \mu_k} = 2 \sum_{i=1}^{m} w_{ik} \left( x^i - \mu_k \right) \tag{3.4}$$

$$\Rightarrow \mu_k = \frac{\sum_{i=1}^{m} w_{ik} x^i}{\sum_{i=1}^{m} w_{ik}} \tag{3.5}$$

I hope you now understand why we are determining centroids in order to define the cluster centers. It actually gives us meaningful results now.

## 3.4 Result and Analysis

### 3.4.1 Categories – Types of Restaurants

We begin our analysis by looking at the various categories of venues that exist in Mumbai. As there are many restaurants, we believe that the majority of venues would be restaurants.

From Figure 3.5, we see that the majority of venues are actually Indian Restaurants. This is closely followed by Pizza Places. As you can see, the graph is colored in a VIBGYOR pattern, where violet shows the most popular categories and red shows the least popular ones. For someone who is visiting Mumbai and loves Indian restaurants, they would surely love their stay.

### 3.4.2 Average Rating for the Places

Next, we will explore the ratings of various venues in Mumbai. We decided to plot a bar chart with x-axis as the rating from 1 to 5 and the y-axis as the count of venues with that rating. We decide to plot the bar chart to see what average rating venues get in Mumbai. This can be seen in 67.

The plot is colored in a VIBGYOR pattern as there was in the previous figure. Here, the violet shows the number of venues with low ratings (around 3) while red shows the number of venues with a high rating (around 5). While the whole range of rating of venues

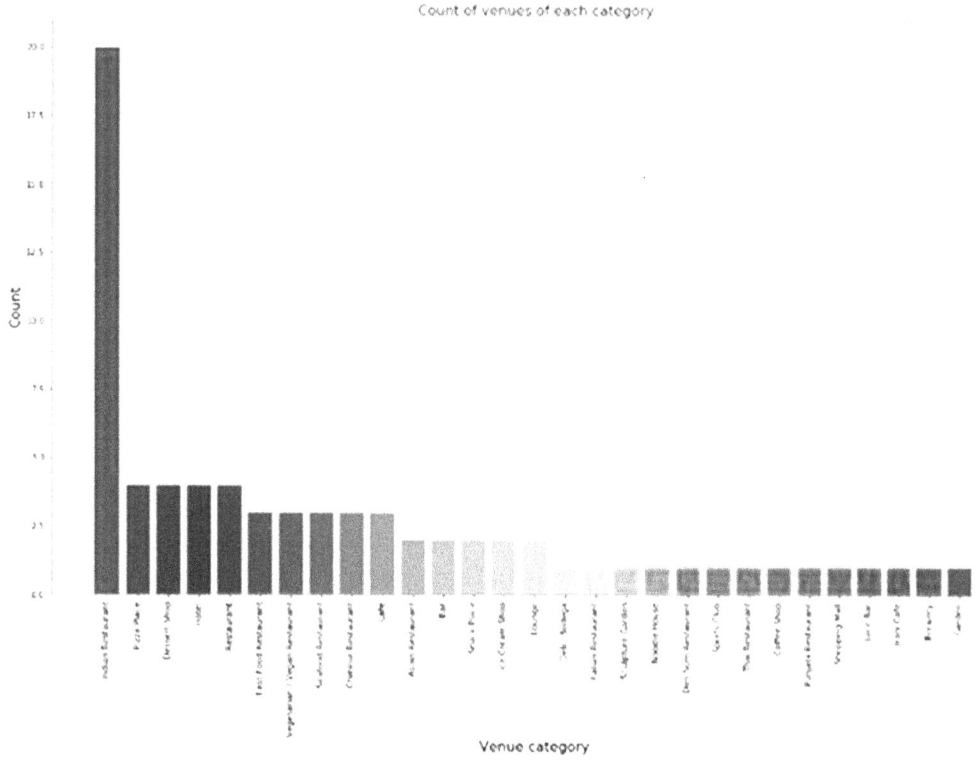

**FIGURE 3.5**
Count of various types of venues in Mumbai.

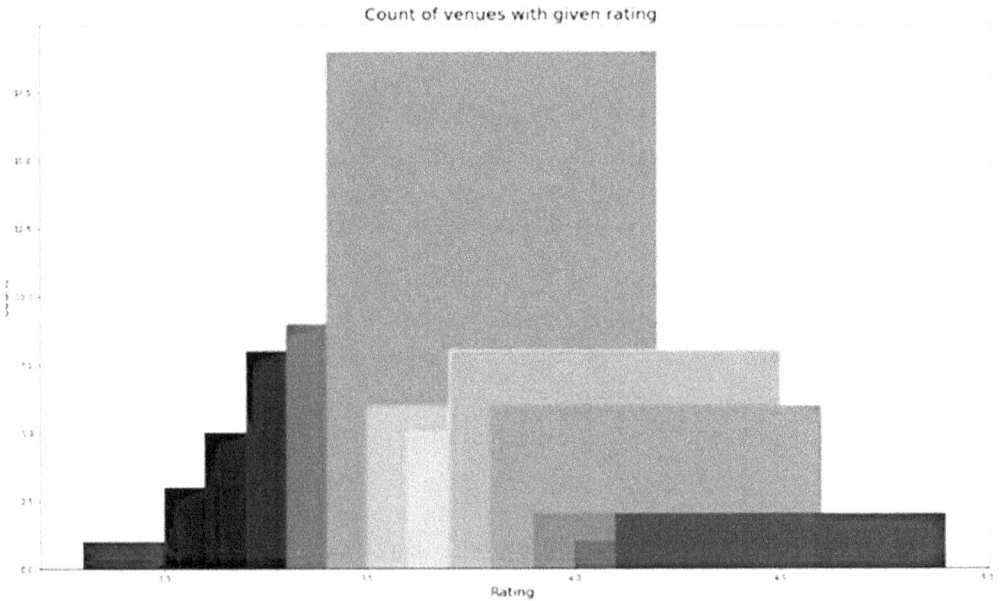

**FIGURE 3.6**
Rating and count of venues with that rating.

might stretch from 1 to 5, the average rating is spread across 3.75 with the maximum number of venues scoring between 3 and 5.

We follow this information by plotting the venues on the map of Mumbai. The venues that are rated below 3 were marked by red and orange while the venues that are rated more than or equal to 3 are plotted as green and dark green. Looking at Figure 3.7 reveals the same results as the bar plot. However, it is interesting to note that many high-rated venues are located near Bandra Complex, and Marol Naka. Ghatkopar has venues with ratings in the complete range from low to high. In addition, the belt of venues in Chembur have high-rated venues. The venues in Mumbai that do not have many venues have ratings of more than three. Overall, Mumbai on average has a good rating for its venues.

### 3.4.3 Average Price Per Person of Venues

Next, we explore the average prices of all venues for one person using a scatter plot along with the count of venues with that average price per person. Taking a look at Figure 3.8, reveals that the majority of venues have an average cost of Rs. 500 to Rs. 1000 for one person. This graph is also colored in a VIBGYOR pattern. Here, violet shows the count of venues having a low price, while red shows the count of expensive venues. Even though the maximum venues lie in that range, the actual range of prices is very different. There are places with an average price even as high as Rs. 1000+ for one person.

We will now plot our data on the map. We see markers with different colors, which represent the price ranges. The red markers are the places that have the higher range of average prices. The yellow ones have medium range and green markers have a low range as seen in Figure 3.9.

Now, we will discuss our implementation of the K-means algorithm.

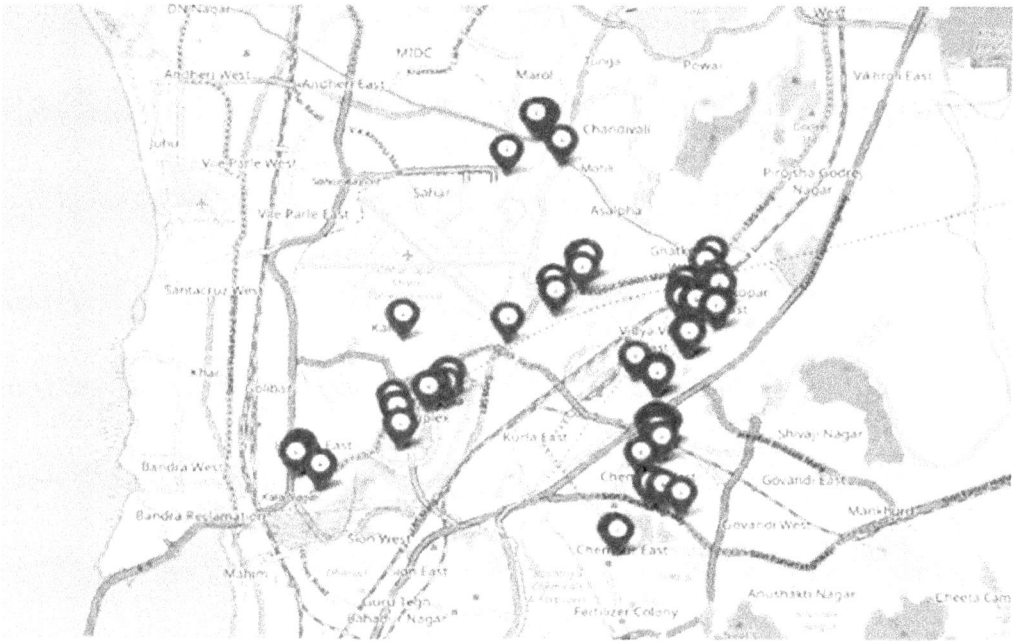

**FIGURE 3.7**
Plot of venues with different ratings.

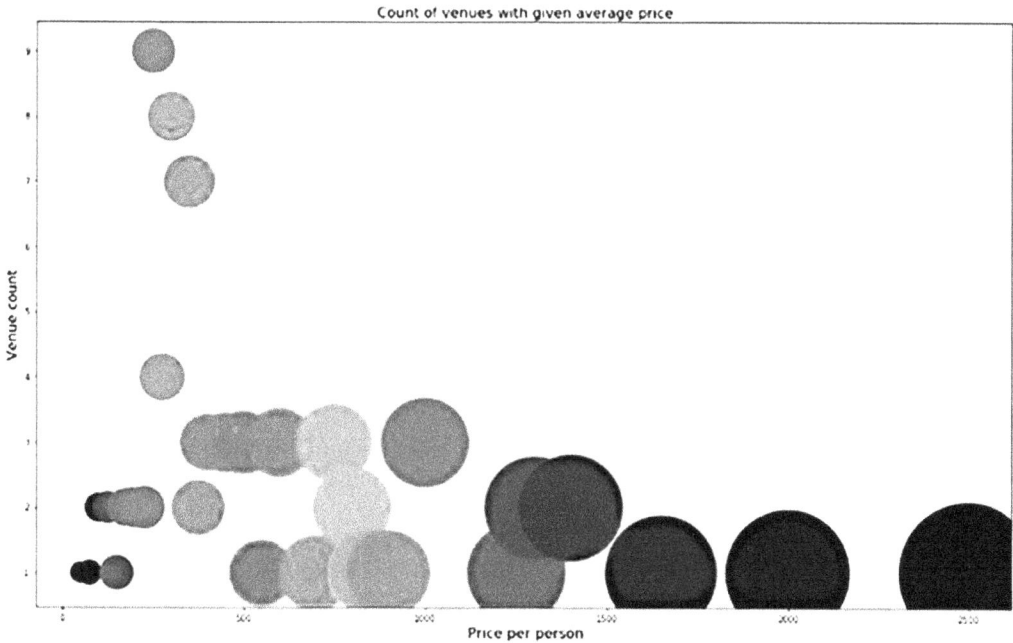

**FIGURE 3.8**
Price per person with a count of venues with that price.

**FIGURE 3.9**
Plot of venues with different prices.

Finally, we cluster all the venues based on their price range, location, and more to identify similar venues and the relationship among them. We used K-means clustering and decided to cluster the venues into two separate groups. In Figure 3.10, we see the two clusters: the first cluster (red) is spread across the whole city and includes the majority of venues. These venues have a mean price range of 2.02 and a rating spread around 3.79. The second cluster (green) is very sparsely spread and has very limited venues. These venues have a mean price range of 3.92 and a rating spread around 4.17. After aggregating data from Foursquare and Zomato APIs, we get a list of 181 different venues. However, not all venues from the two APIs are identical. Hence, we have to inspect their latitude and longitude values as well as names to combine them and remove all the outliers. This will result in a total venue count of 75. We identified that from the total set of venues, the majority of them are Indian restaurants. A visitor who loves Indian Restaurants would surely like to visit Mumbai. While the complete range of ratings ranges from 1 to 5, the majority of venues have ratings close to 3.75. This means that most restaurants provide good quality food, which is liked by the people of the city, thus indicating the high rating. When we look at the price values of each venue, we explore that many venues have prices, which are in the range of Rs. 500 to Rs. 1000 for one person. However, the variation in prices is very large, given the complete range starts from Rs. 100 and goes to Rs. 2500. On plotting the venues based on their price range on the map, we discovered that venues located near Ghatkopar and Chembur are relatively priced lower than venues in Bandra Complex and Marol Naka. Finally, through clusters, we identify that there are many venues, which are relatively lower priced but have an average rating of 3.79. On the other hand, there are few venues which are high priced and have an average rating of 4.17. If you are looking for cheap places with a relatively high rating, you should check Ghatkopar and Chembur. If you are looking for the best places, with the highest rating but might also carry a high price tag,

**FIGURE 3.10**
Clusters of venues.

you should visit Bandra Complex. A company can use this information to build up an online website/mobile application, to provide users with up-to-date information about various venues in the city based on the search criteria (name, rating, and price) or to start a food delivery business.

## 3.5 Conclusion

Data science is the hottest field in the world. It has been said that data scientists are to this century what computer programmers were to the last century. Yet, data science is a relatively new field, and most people know very little about it. The Scope of Data Science is a comprehensive guide to data science and the many disciplines to which it relates.Data Science has been evolving rapidly in the last few years. It is now a hot field with a high demand for data science professionals. Data Science is not only about analytics and statistics, but also about machine learning and data engineering, and there are more and more tools and libraries to do these tasks.

The purpose of this chapter was to explore the places that a person visiting Mumbai could visit. The venues have been examined using Foursquare and Zomato API and have been plotted on the map. The map reveals that there are three major areas a person can visit: Ghatkopar, Bandra Complex, and Chembur. Based on the visitor's venue rating and price requirements, he/she can choose among the three places. In this chapter, the complete analysis is done using python libraries on Jupyter Notebook. Using the APIs like Zomato and Foursquare, the analysis could be applied to any location in the world. We would also like to continue future development of this project by deploying a web application for the same analysis so that it would help the visitors in real-life scenarios.

---

# References

1. Krueger, Alan B., and Mikael Lindahl. "Education for growth: Why and for whom?." *Journal of Economic Literature* 39, no. 4 (2001): 1101–1136.
2. Davenport, Thomas H., Jeanne G. Harris, David W. De Long, and Alvin L. Jacobson. "Data to knowledge to results: Building an analytic capability." *California Management Review* 43, no. 2 (2001): 117–138.
3. Gullo, Francesco. "From patterns in data to knowledge discovery: What data mining can do." *Physics Procedia* 62 (2015): 18–22.
4. Schuster, Alfons J. "Intelligent computing everywhere." In *Intelligent computing everywhere*, pp. 3–23. Springer, London, 2007.
5. Yang, Qiang, Yang Liu, Tianjian Chen, and Yongxin Tong. "Federated machine learning: Concept and applications." *ACM Transactions on Intelligent Systems and Technology (TIST)* 10, no. 2 (2019): 1–19.
6. Unold, O., and L. Cielecki. "Learning Context-Free Grammars from Partially Structured Examples: Juxtaposition of GCS with TBL." In: *7th International Conference on Hybrid Intelligent Systems (HIS 2007)*, Kaiserlautern, 2007, pp. 348–352. doi:10.1109/HIS.2007.44.
7. Anderson, John Robert. *Machine learning: An artificial intelligence approach*. Vol. 3. Morgan Kaufmann, San Mateo, CA, 1990.
8. Li, H., S. Fang, S. Mukhopadhyay, A. J. Saykin, and L. Shen, "Interactive Machine Learning by Visualization: A Small Data Solution." In: *IEEE International Conference on Big Data (Big Data)*, Seattle, WA, USA, 2018, pp. 3513–3521. doi:10.1109/BigData.2018.8621952.
9. Kalra, Vaishali, and Rashmi Aggarwal. "Importance of Text Data Preprocessing & Implementation in RapidMiner." In: *The First International Conference on Information Technology and Knowledge Management*, 2017, pp. 71–75. doi:10.15439/2018KM46
10. Cox, Nicholas, and Kelvyn Jones. *Exploratory data analysis: Quantitative geography*. Routledge, London, 1981, pp. 135–143.
11. Chris, Chatfield, "Exploratory data analysis", *European Journal of Operational Research* 23 (1986), 5–13.
12. Greiff, W.R. The use of Exploratory Data Analysis in Information Retrieval Research. In: Croft W.B. (eds) *Advances in Information Retrieval: The Information Retrieval Series*, vol. 7. Springer, Boston, MA, 2002. doi:10.1007/0-306-47019-5-2.
13. Li, Y. et al. "Exploring venue popularity in Foursquare". In: *2013 Proceedings IEEE INFOCOM*. 2013, pp. 3357–3362. doi:10.1109/INFCOM.2013.6567164.
14. Bishnu, P. S. and V. Bhattacherjee. "Software Fault Prediction Using Quad Tree-Based K-Means Clustering Algorithm". IEEE Trans-actions on Knowledge and Data Engineering 24.6 (2012): 1146–1150. doi:10.1109/TKDE.2011.163.
15. Badhiye, S. S., P. N. Chatur, and B. V. Wakode. "Temperature and humidity data analysis for future value prediction using clustering technique: An approach." *International Journal of Emerging Technology and Advanced Engineering* 2.1 (2012): 88–91.
16. Durairaj, M., and C. Vijitha. "Educational data mining for prediction of student performance using clustering algorithms". *International Journal of Computer Science and Information Technologies* 5.4 (2014): 5987–5991.
17. Sivaranjani, S., Sivakumari, S., and Aasha, M. "Crime prediction and forecasting in Tamilnadu using clustering approaches". In: *2016 International Conference on Emerging Technological Trends (ICETT)*. 2016, pp. 1–6. doi:10.1109/ICETT.2016.7873764.23.

18. Gan, S. et al. "Ship trajectory prediction for intelligent traffic management using clustering and ANN". In: *2016 UKACC 11th International Conference on Control (CONTROL)*. 2016, pp. 1–6. doi:10.1109/CONTROL.2016.7737569.

19. G. Pettet et al. "Incident analysis and prediction using clustering and Bayesian network". In: *2017 IEEE Smart World, Ubiquitous Intelligence Computing, Advanced Trusted Computed, Scalable Computing Communications, Cloud Big Data Computing, Internet of People and Smart City Innovation (Smart World/SCALCOM/UIC/ATC/CBDCom/IOP/SCI)*. 2017, pp. 1–8. doi:10.1109/UIC-ATC.2017.8397587.

20. Metfessel, Brent A. et al. "Cross-validation of protein structural class prediction using statistical clustering and neural networks". *Protein Science*, 2.7 (1993): 1171–1182.

21. Tharavath, V., Gupta, D., and Gunasekar, S. "Factors influencing choice of cuisines while Indian consumers eat out." In: *2017 International Conference on Data Management, Analytics and Innovation (ICDMAI)*, Pune, 2017, pp. 157–160. doi:10.1109/ICDMAI.2017.8073502.

22. Foursquare Dataset for food venues, Foursquare API: https://developers.foursquare.com/api.

23. Zomato Dataset for Restaurants, Zomato API: https://developers.zomato.com/api

24. Garima, H. Gulati and P. K. Singh, "Clustering techniques in data mining: A comparison." In: *2015 2nd International Conference on Computing for Sustainable Global Development (INDIACom)*, New Delhi, 2015, pp. 410–415.

25. Bindra, K. and Mishra, A. "A detailed study of clustering algorithms." In: *2017 6th International Conference on Reliability, Infocom Technologies and Optimization (Trends and Future Directions) (ICRITO)*, Noida, 2017, pp. 371–376. doi:10.1109/ICRITO.2017.8342454.

26. Ahalya, G. and Pandey, H. M. "Data clustering approaches survey and analysis." In: *2015 International Conference on Futuristic Trends on Computational Analysis and Knowledge Management (ABLAZE)*, Noida, 2015, pp. 532–537. doi:10.1109/ABLAZE.2015.7154919.

27. Oyelade, J. et al., "Data Clustering: Algorithms and Its Applications." In: *19th International Conference on Computational Science and Its Applications (ICCSA)*, Saint Petersburg, Russia, 2019, pp. 71–81.

28. Rui, Xu and D. Wunsch, "Survey of clustering algorithms." *IEEE Transactions on Neural Networks*, 16, no. 3 (May 2005): 645–678. doi:10.1109/TNN.2005.845141.

29. Wilkin, G. A. and X. Huang, "K-Means Clustering Algorithms: Implementation and Comparison." In: *Second International Multi-Symposiums on Computer and Computational Sciences (IMSCCS)*, Iowa City, IA, 2007, pp. 133–136. doi:10.1109/IMSCCS.2007.51.

30. K-Means Clustering: Algorithm, Applications, Evaluation Methods, and Drawbacks. https://towardsdatascience.com/k-means-clustering-algorithm-applications-evaluation-methods-and-drawbacks-aa03e644b48a

# 4

## Investigation on Mobile Forensics Tools to Decode Cyber Crime

Keshav Kaushik

*School of Computer Science, University of Petroleum and Energy Studies, Dehradun, Uttarakhand, India*

## CONTENTS

## 4.1 Introduction to Mobile Forensics: Process and Methodology

Mobile Forensics is indeed very challenging, all of it just boils down to one simple factor, and that is the lack of standardization present. Not to mention other possible factors such as limited processing power and memory resources (though that does not seem to be the case as flagship smartphones continuously evolve into standing their own against even some budget laptops. But this is exactly where standardization comes in where you just have huge range of mobile phones to choose from and use), unique processors architectures, and the range of operating systems available in the market. Mobile forensics in some contexts can be divided into three categories: acquired data types, operating systems, and evidence acquisition methods. Coming to acquisition methods, we can further divide them into three types, namely physical, manual, logical. Manual acquisition is the process in which an individual acquires information while interacting with the divide itself. This procedure has to perform while keeping a log of all the actions performed as human errors have a high risk, while the individual interacts with installed applications to copy existing data. This method is best used as supplementary since it is also the only method in which the data collected is in a human-readable format.

Then we have logical acquisition, which collects the bit-by-bit clone of directories and files stored in the logical partition. Since we know the information is never practically deleted but overwritten, we have the physical acquisition method come in which helps in retrieving the deleted files. There will be times when logical acquisition would not be possible to a damaged device, which is another reason for physical acquisition to step in. As

DOI: 10.1201/9781003206088-4

mobile devices also continue to grow, becoming faster and more reliable we also have traditional crimes being mobile concentric as the popularity with devices keeps on increasing year by year. At present, mobile forensics is dependent upon the tools and techniques pertaining to mobile phone vendors. The chapter will discuss the role of mobile forensics in cybercrime investigation and security analytics. The various applications and challenges of the mobile forensics process are also highlighted in this chapter. It also explores the tools and techniques related to mobile forensics. The book chapter also highlights the managing of digital evidence by maintaining the chain of custody.

The Internet of Things (IoT), cloud technology, and data analytics are all burgeoning technological advancements, and mobile phones are smack in the center of them all. The growth of mobile technology is arguably the most important cause for these developments to emerge in the first position, or at certainly one of the most important ones. In 2015 (*The Mobile Forensics Process: Steps and Types – Infosec Resources*, 2019), the United States had 377.9 million wireless customer accounts, including feature phones, tablets, and smartphones. Currently, the smartphone application is as ubiquitous as it is helpful, especially in the context of digital forensics, because these small gadgets gather huge volumes of data on a daily basis, that may be recovered to assist the examination. The objective of forensic investigation is to recover forensic evidence or relevant data from a smartphone while keeping the data exhaustively admissible. To do this, the mobile forensic method must create precise rules for securely capturing, segregating, transferring, preserving for investigation, and certifying digital evidence obtained through cell devices. Mobile forensics techniques are typically equivalent to those used in other forensic science fields.

The process of mobile forensics is shown in Figure 4.1. As per computer forensics, the evidence must always be properly stored, examined, and recognized in a legal proceeding. Mobile device seizures (EclipseForensics, 2021) are followed by a slew of legal concerns. It's usually a good idea to isolate the link, which you can accomplish by utilizing between (1) Airplane Mode + Barring Wi-Fi and Hotspots, or (2) duplicating the phone Subscriber Identity Module (SIM) card. So the aim of their capture is to gather evidence, the simplest method of transferring them is to maintain devices switched on in order to prevent discontinuation, which would undoubtedly change information. An external gadget and a Faraday bag are standard pieces of mobile forensics gear. The first is a box meant to separate mobile devices from wireless networks while also assisting in the secure transfer of captured evidences to the forensic lab, while the latter is a power supply integrated within the Faraday bag. Finally, detectives ought to be wary of portable devices linked to unidentified gadgets, and other tripwire put up to implement physical damage or death to anybody present at the area of the crime (Figure 4.2).

The data collection phase's goal is to obtain information from the cell device. A closed display can only be unlocked with the right PIN, password, pattern, or fingerprints. Because data on portable devices is moveable, it is tough to keep track of it. Whenever texts or data are moved from a mobile, access is jeopardized. Regardless of the fact that many gadgets are capable of storing enormous amounts of data, the information itself may be stored somewhere else. Forensic experts should be on the lookout for any signs that data is escaping the mobile phone and becoming a physical item since this might have an impact on the collecting and even preservation process. Enable the collection of data resources, the data must be appropriately gathered. Whenever it comes to obtaining info, there are some unique issues that have arisen as a result of the effect of computer technology. Several mobile devices cannot be gathered simply producing a picture, therefore they'll have to go through a data-gathering method. Because certain design constraints may only permit one form of collection, there are a variety of techniques for acquiring data from mobile devices.

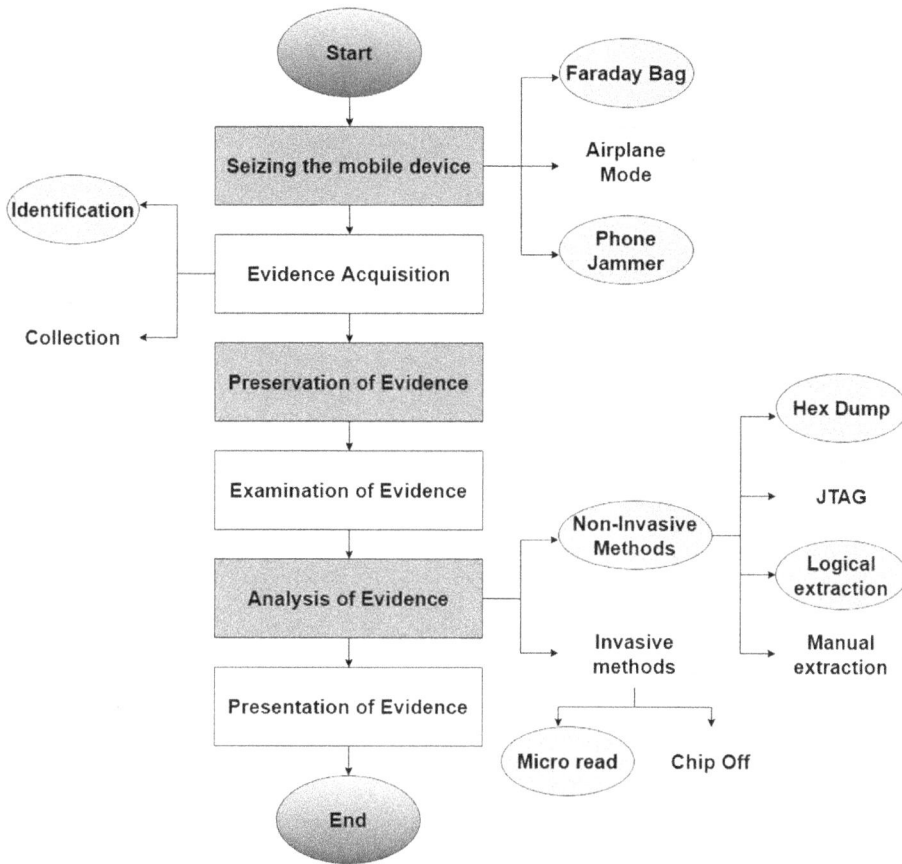

**FIGURE 4.1**
A flowchart showing the mobile forensics process diagram, with six steps starting from seizing the mobile device and presentation of evidence as the final step.

The examiner might need to use a number of forensic techniques to get and assess data contained in the database. Because of the wide variety of portable devices, there really is no one-size-fits-all method for mobile forensic tools.

The evidence extraction techniques are as follows:

- Manual: A method of documenting data stored in the phone's storage by manually using the keyboard and phone screen.
- Logical: A method for extracting certain files or entities.
- File System: A process that extracts files from a file system and may include data marked for deletion.
- Physical (Invasive): A method for physically acquiring data from a device that necessitates dismantling of the gadget to get contact to the circuit board.
- Physical (Non-Invasive): A method for physically acquiring data from a device without having to open the phone's casing.
- Chip-Off: A disruptive operation in which a memory chip is removed and read in order to undertake the examination.

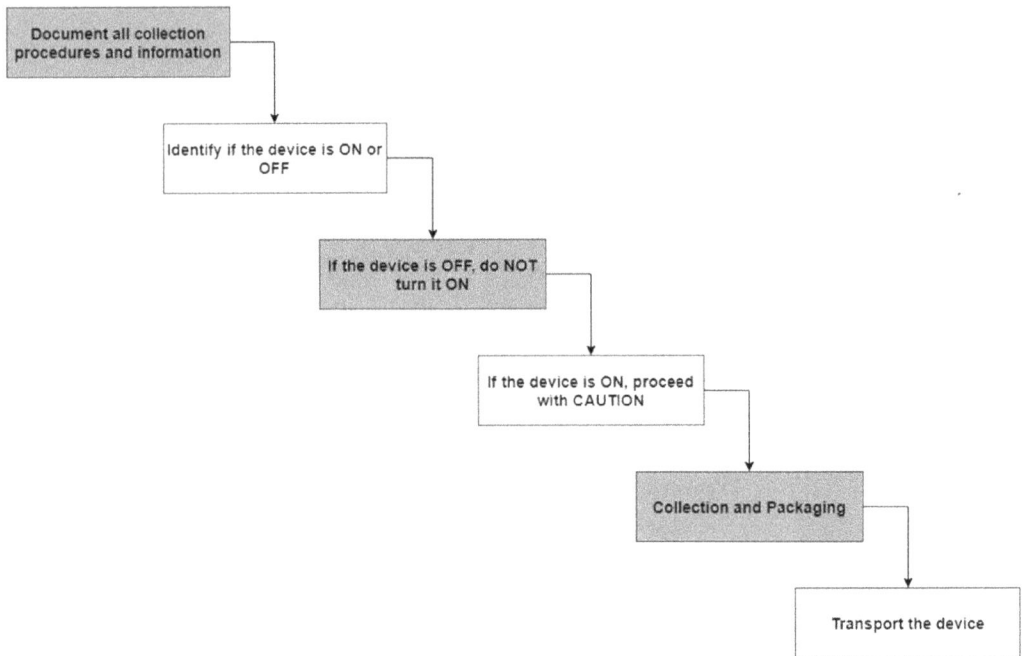

**FIGURE 4.2**
A step-by-step diagram showing the various stages of the mobile acquisition process.

- ISP (In-System Programming): A method of dismantling a device and attaching it to memory chip pinouts.
- JTAG (Joint Test Action Group): A method of dismantling a gadget and attaching it to check access ports.

Mobile device forensics is a field in which various different methods and techniques are implemented on a massive variety of mobile gadgets but not limited to smartphones. Usage of information technology and the internet are no longer a luxury but rather a part of our daily life and smartphones becoming more common. They possess the ability to store a lot of personnel information data on its user not to mention other running location-based, internet services, that really do make a smartphone an ideal candidate to perform investigation upon. A significant amount of research has taken place over the years on various aspects of mobile forensics including various data acquisition schemes, information methods, and various mobile device platforms. This lack of standardization among the smartphone manufacturers leads us to the scarcity of interfaces, hardware and software, and standardized guidelines of the mobile forensics industry. Though luckily, we have had many researchers pour in their work that has led to creating some tools and guidelines like APCO (The Association of Chief Police Officers), NIST (The National Institute of Standards Technology).

Mobile forensics in some context can be divided into three categories: acquired data types; evidence acquisition methods; operating systems. Regarding acquisition methods, we can further divide them into another three types namely, physical, logical, and manual. The manual acquisition is the process in which an individual acquires information while interacting with the divide itself. This procedure has to perform while keeping a log of all the actions performed as human errors have a high risk, while the individual interacts with installed applications to copy existing data. This method is best used as supplementary since it also the only method in which the data collected is in a human-readable format.

Despite having so many tools for hard disk drive analysis, making digital copies, decrypting files, recovering passwords, etc. Although quickly improving, the technologies we use in forensic investigation, such as Micro Systemation's XRY, Paraben's Device Seizure, and BitPim, have limits. There are so many mobile phones in the market each incorporating different features, a different and or newer version operating system, creating and managing support is truly very tasking. According to DeepSpar (*DeepSpar Data Recovery Systems: Data Recovery Equipment, Hardware, Business Solutions.*, n.d.) from Data Recovery, 50% of hard disk failures are caused by faulty firmware, which may be fixed by reinstalling the firmware. The problem with mobile devices is that there are many more layers into play than just the firmware and hence firmware has to be flashed which leads to overwriting of all the existing data.

Despite all these physical acquisition methods, the most common is the use of modified bootloaders; regardless of the target device's operating system, installing your own modified bootloader will almost always work. Maturing of the field of mobile forensics has led to low-level modifications in the mobile operating systems as to exploit vulnerabilities of the operating system, which would make evidence at times inadmissible in court. These lower-level modifications grant access to system areas that were locked down by the manufacturers.

*Rooting on Android, jailbreaking in iOS, or capability hack in Symbian OS*: Now, coming to the data types that can be acquired, the first group can be the data handled by the operating system; for example, GPS and Wi-Fi connection managers, as well as structural components like international mobile equipment identity (IMEI) and international mobile subscriber identity (IMSI). The second category includes data that users have imported and changed, such as messages, phone numbers, photos, and other customized application programs.

The ACPO standards (Owen & Thomas, 2011) do not provide suggestions for all forensic hardware or software that may be needed during an investigation, but NIST takes into account all of the devices that should be acquired, including software and hardware, to guarantee that forensic standards are maintained. ACPO, on the whole, lacks a standardization that can be sustained between probes. Though the ACPO recommendations include legal implications and standards to follow in order to maintain evidence integrity, the NIST recommendations are not intended to provide legal advice and are instead utilized as a reference point. The ACPO rules, on the other hand, do not specify how criminal justice system should handle mobile devices throughout an examination.

## 4.2 Preservation of Evidences in Mobile Device

The discovery, identification, documenting, and collecting of electronic-based data are all part of the preservation process (Ayers et al., 2014). Evidence must be maintained in order to be used effectively, in either the less formal procedure or a law court. Inability to uphold evidence in its original condition might endanger a whole investigation, resulting in the loss of critical case evidence. When safeguarding a smartphone, caution should be exercised when allowing a person to utilize the gadget. Several mobile devices include master restore passwords that restore the device's data to factory settings. It is advisable (but may not be practical) to acquire data from a mobile device using two or more methods (i.e. comparing results). Insider attacks (Kaushik, 2021) plays a major component of cyber-attacks. A manual examination is sometimes okay if investigators only need a particular piece of information and using an identical test device beforehand. Most labs that deal with mobile devices keep a large collection. Forensic examination and analysis of mobile data are done to find and extract information related to an investigation (including deleted data). Generic phases (*Digital Evidence Preservation – Digital Forensics – GeeksforGeeks*, 2020) in the preservation of mobile evidence are given below:

- Do not modify the phone's present state: if it's off, it should stay that way; if it's on, it should stay that way. When you do something, consult a forensics specialist.
- Turn off the gadget: Throughout the case of mobile devices, do not charge them if they are not fully charged. If the phone is turned on, turn it off to avoid information erasure or rewriting due to automated rebooting.
- Do not keep the gadget unsupervised in an exposed or insecure location: Do not leave the gadget unsupervised in an outdoor or unprotected location. One must keep track of where the gadget is located, who has permission to it, and when it is transferred.
- No external memory devices should be plugged into the gadget: Memory sticks, USB flash drives, and other types of storage devices must not be connected to the machine.
- Nothing should be copied to or from the gadget: When you copy something to or from the gadget, the slack space in the memory changes.
- Snap a picture of the evidentiary component: Make careful to photograph the evidence from all angles. If it's a cellphone, take photographs from every angle to confirm the gadget hasn't been tampered with until forensic experts arrive.
- Remember to check the phone's PIN/password pattern: It is critical that you understand the device's username and password and provide these to the forensic specialists so that they can complete their work efficiently.
- On the phone, do not access any images, apps, or files: Any program, file, or image opened on the gadget may result in data loss or ram overwriting.
- Do not put your faith in someone who hasn't had forensics education: The files on the actual phone should only be investigated or viewed by a trained forensics specialist. Untrained individuals may lead data to be deleted or vital information to be corrupted.
- Ensure the computer is not turned off, and hibernate it if necessary: Because digital evidence can be retrieved from both hard drives and volatile storage, it is a viable option. The information of volatile memory will be preserved in hibernation mode until the next system startup.

## 4.3 Acquisition from Mobile Device

Acquisition is the process of scanning or otherwise gathering information from a mobile telephone and its associated media. Conducting a collection on-the-spot offers the benefit of preventing data loss due to battery exhaustion, breakage, and other factors during storage and transportation. In contrast to a laboratory environment, obtaining a controlled environment with which to operate with the required equipment while meeting extra requirements might be difficult in off-site collections. The recognition of the mobile device is the first step in forensic investigation. The method to pursue in generating a forensic replica of the information of a mobile device is determined by the device type, its features, and the operating system. The tools and procedures that should be utilized in an inquiry depend on the type of mobile phone and information to be retrieved. The generic mobile acquisition process is given below.

- In step 1, all collection procedures and information are well documented with the help of notes, photograph/video/sketch, and chain of custody is maintained.

- In the next step, the forensic experts check whether the device is ON or OFF. This can be done by observing the lights in the phone, and by feeling the vibration or listening to the sounds.

- After confirming that the device is OFF, then do NOT turn it ON. Forensic experts should ask for some passwords or patterns and directly proceeds to transport after packaging.

- The devices should be handled carefully if it is switched ON. We should isolate the device from Wi-Fi or cellular networks. In addition, we should try to obtain security passwords or PIN.

- In this step, we should collect all the evidence and do the packaging properly after labeling.

- Finally, the mobile evidence is carefully transferred as quickly as possible to a guarded police department's station or a digital forensic facility. The transportable evidence should be safeguarded from excessive moisture and temperature while being transported.

## 4.4 Examination and Analysis: Tools and Techniques

Rapidly increasing in the size of pictures and videos nowadays, the upcoming mobile devices are having a larger amount of storage (128 GB +) and it is making it difficult for investigators to investigate complete devices because the higher the amount of data, the higher the time it takes. Therefore, the research done in (Guido et al., 2016) is there to decrease the overall acquisition time. Various security tools (Kaushik et al., 2020) are helpful in digital forensics and for the detection of cybercrime. Security tools like autopsy are very helpful in analyzing the evidences and performing the forensic investigation by maintaining the chain of custody.

For that purpose, developers make a tool by doing prior research name it as Hawkeye. It is a prototype and it works on different analyses and runs within the Android custom bootloader. Hawkeye is a simple C Program compiled for the arm, which is temporarily stored over random access memory of the device. This tool is made to run at one time whenever the acquisition is required at a crime scene. After setting up everything at the crime scene the tool provides two information: (i) Base Line Hash List and (ii) Partition Data perceived from perusing and harvesting the device's GUID Partition Table (GPT). Partition data is noticeable in path and size format, which would be Comma Segregated. There are a variety of data collection methodologies and evidence extraction process paradigms and standards available today. However, when combined with the continuously changing technological world, this heterogeneity adds to the forensic investigative process's complexity. Researchers (Chernyshev et al., 2017) provide a detailed overview of current state-of-the-art phone forensic procedures and highlight research possibilities that must be pursued in the future to allow performance that is more effective.

SEAKER (Gentry & Soltys, 2019) stands for Storage Evaluator and Knowledge Extraction Reader, and it aids in the rapid triage of a large number of digital gadgets. Designed for on-the-spot, time-sensitive assessments, its usefulness attains out to avoid over-assortment and massive abuses at computerized legitimate science research labs around the world. IoT gadgets are present in every regular family nowadays, things like pro-cameras, refrigerators, thermostats, light bulbs, window covers. The enormous sum of computerized data that is being produced by these devices is gathered which is frequently useful in criminal examinations. Standard HTML is used as the investigator's unique identifier (UI), taking into account with greatest similarity with numerous different searching gadgets through the internet browser.

Because the industry is inundated with both open source and commercial mobile phone operating platforms, the approaches and tools now available fail to obtain comprehensive information from the gadgets, and selecting the right tool is difficult. This research (Sathe & Dongre, 2018) examines the physical and logical collection strategies employed by forensic investigators, and so gives a comparative analysis of which methodology is more effective in obtaining digital evidence from cell devices. Smartphones are an essential part of everybody's lives, which has resulted in illegal activities such as phishing, Spear phishing, and SMS faking. Criminals have tried to erase data from digital evidence in a cellular telephone. Forensic experts can learn about users' data by using data from mobile devices. A unique approach for data gathering of forensic evidence from a hacked device is presented in this article (Maria Jones et al., 2017) that will be beneficial for digital investigators and court processes. This type of data collection allows for a better understanding of the offenders who employed mobile devices for cyber-attacks.

The mobile forensic field is compared in terms of the Android forensic method for gathering and analyzing an Android disk image. In this article (Aiman & Ernest, 2019), the problems of Android forensics are characterized, such as the sophistication of the Android application, various methodologies and systems for analyzing information, problems with hardware configuration, and the use of expensive business tools for procuring logical data that struggle to fetch physical data acquisition. A platform-independent approach is being developed to overcome these issues and achieve high authenticity and precision in Android forensic operations. The mobile forensics (Barmpatsalou et al., 2018) field was born out of the requirement to investigate the above-mentioned occurrences. A sub-domain of digital forensics, mobile forensics specializes in collecting and analyzing data from smartphones in order to identify and trace attacker organizations and operations. Mobile forensics has lately broadened its focus to include the structured and sophisticated depiction and evaluation of potential hostile entity activity, in addition to its core research emphasis in evidence gathering from smartphones. Despite this, data collection is still the company's primary priority.

With advancements in technology, operating capability, processing power, and usefulness, the necessity of evaluating data termed evidence in mobile phones has grown. Authorized personnel in a forensics investigation must check cellphones, and the data collected from the gadget must be brought up to forensic criteria. Examining and analyzing mobile phones in context of online forensics is examined in this study (Dogan & Akbal, 2017). Simultaneously, data received from mobile phones using a sample project has been analyzed. This paper (Kumar et al., 2018) report about the forensics approach toward mobile application malware. According on the study's findings, a balanced scoring rate list of characteristics was created, which was then used to create a proactive and compact malware detection for Android phones. This developed model was used to create FAMOUS (Forensic Analysis of Mobile Devices Using Scoring of Application Permissions), an investigative tool that can screen all of a linked device's installed applications and generate a comprehensive summary.

The capability of XRY (Ayers & Rick, 2020) to get relevant data from the internal storage of based mobile devices and related media was evaluated. Excluding the aforementioned abnormalities, the program correctly and thoroughly collected all available data items for all portable devices evaluated. Smartphones have progressed in a range of methods in recent years, including such intrinsic power source capability, internal memory processing, and CPU functionality, allowing owners of smartphones to increase processing power while preserving a handheld size, effectively turning it into a compact storage device where people carry their private information. Because of these developments in the nature and utility of smartphones, they have become more important in areas like legal consequences in police or corporate examinations. Researchers will implement a systematic review of the topic of forensic investigation in this article (Gill et al., 2018), which will allow us to analyze the level to which this major innovation can become relevant evidence. The breakthroughs forensic experts have made over time on the subject, the probable innovations that could control more developments in the process of mobile forensics and its influence, as well as the differences between the two.

## 4.5 SIM Forensics and Reporting of Evidences

A SIM card has a CPU and operating system, as well as 16–256 KB of read-only storage that is permanent, electrically erasable, and programmable (EEPROM). It also has RAM (random access memory) and ROM (read-only memory). USIMs are upgraded variants of today's SIMs that carry information that is backward compatible. A USIM is distinct in that it permits a single phone to have numerous contact information. Cyber forensic science is a forensic professional's ability to use computer science expertise and investigation techniques in a legal case that requires the examination of forensic evidence. It is the procedure for locating, conserving, evaluating, and presenting digital evidence in a legal-acceptable manner. The goal of the procedure (Srivastava & Vatsal, 2016) is to retain any forensic evidence in its most original state while doing a planned study that includes finding, gathering, and verifying digital data in order to recreate previous occurrences.

SIM cards are interchangeable with smartphones that use GSM mobile networks. In the GSM framework, a mobile device is known as a mobile terminal and is separated into two parts: the universal integrated circuit card (UICC) and the mobile equipment (ME). A UICC is a detachable component that holds vital client information. It is also known as an identity module (e.g., Subscriber Identity Module [SIM], CDMA Subscriber Identity Module [CSIM], Universal Subscriber Identity Module [USIM]). A UICC is required for the ME and wireless phone to work properly. There are three distinct sizes of UICCs (Ayers et al., 2014) provided as shown in Figure 4.3. Mini-SIM (2FF), Micro SIM (3FF), and Nano-SIM are the three options (4FF). With a width of 25 mm, a height of 15 mm, and a thickness of 0.76 mm, the Mini-SIM is about the size of a postage stamp and is now the most widely

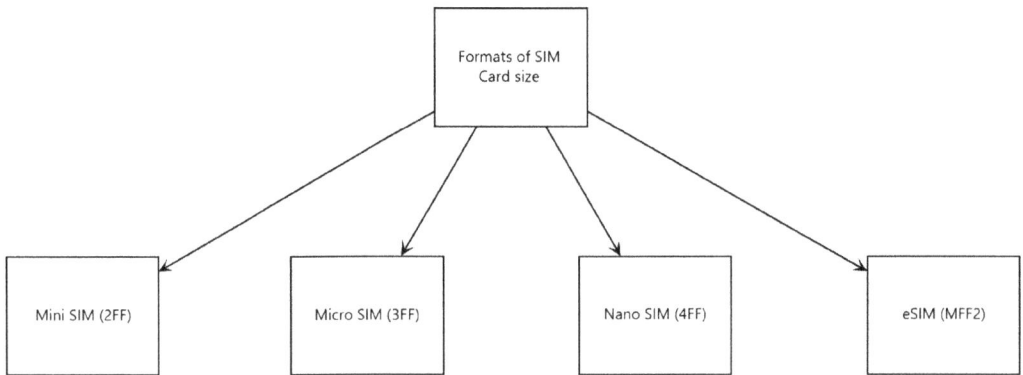

**FIGURE 4.3**
A hierarchical diagram showing the formats of SIM card size

used format on the planet. Modern mobile handsets utilize Micro (12 × 15 ×.76 mm) and Nano (8.8 × 12.3 ×.67 mm) SIMs (e.g., iPhone 5 uses the 4FF).

A UICC can retrieve many forms of digital information from fundamental data files dispersed across the system files. Many of the same information stored in the UICC may also be stored in the mobile device's storage and accessed there. A UICC may include non-standard files created by the service provider in addition to the standard data described in the GSM standards. A small SIM card contains a lot of data; therefore the forensic investigation of the SIM card is carried out separately. A SIM card has some security features like:

- Administrative (ADM): Only when the consumer's card company has met the administrative access criteria can the customer gain access.
- Always: This situation provides for unrestricted access to data.
- Never (NEV): Using the SIM/ME interface to access the file is prohibited.
- Cardholder verification 1 (CHV1): If the user's PIN is successfully verified or if PIN validation is off, this precondition permits access to files.
- Cardholder verification 2 (CHV2): After successfully verifying the user's PIN2 or if the PIN2 verification is disabled, this condition permits access to files.

There are several tools used for SIM Card Forensics, popular ones are mentioned below:

- MOBILedit Forensic: This program can study phones via Bluetooth, IrDA, or cable connections; it can also analyze SIM cards using SIM readers and read erased messages from them.
- Encase Smartphone Examiner: This tool is meant to collect data from smartphones and tablets, such as the iPhone and iPad.
- pySIM: A SIM card administration program capable of generating, modifying, removing, as well as backing up and restoring SIM phonebooks and SMS data.
- SIMpull: SIMpull is a strong tool, a SIM card capture program that allows you to obtain a SIM card's actual details.
- AccessData Mobile Phone Examiner (MPE) Plus: This application provides over 7000 devices, like iOS, Windows Mobile Blackberry, Android, and Chinese handsets, and is available as equipment that includes a SIM card reader and data connections.

## 4.6 Conclusion

As mobile devices also continue to grow, becoming faster and more reliable, we also have traditional crimes being mobile concentric as the popularity with devices keeps on increasing year by year. At the moment, mobile forensics is reliant on methodologies and technologies that are particular to certain vendors. Although recommendations like NIST and ACPO must be revised on a regular basis as the mobile market advances. Notwithstanding this, mobile device testimony is frequently used as a complement, regardless of the fact that the quantity of information that can be gleaned from a phone is growing and developing fast. As a result, it's reasonable to expect that mobile phone information will progressively play a role as the major source of information in forthcoming judicial proceedings. This chapter highlights the mobile forensics process, tools, and techniques. As the number of mobile devices keeps on increasing, there is a need for more advanced mobile forensics tools that will help forensic investigators in solving cybercrimes.

## References

Aiman, A.-S., & Ernest, F. (2019). A comparison study of Android mobile forensics for retrieving files system. *International Journal of Computer Science and Security (IJCSS)*. https://www.researchgate.net/profile/Aiman-Al-Sabaawi/publication/335422366_A_Comparison_Study_of_Android_Mobile_Forensics_for_Retrieving_Files_System/links/5d650855299bf1f70b0f32ba/A-Comparison-Study-of-Android-Mobile-Forensics-for-Retrieving-Files-System.pdf

Ayers, R., Brothers, S., & Jansen, W. (2014). *NIST Special Publication 800-101 Revision 1 Guidelines on Mobile Device Forensics*. doi:10.6028/NIST.SP.800-101r1.

Ayers, R., & Rick. (2020). *Test Results for Mobile Device Acquisition Tool: MSAB XRY v9.0.2*. http://www.cftt.nist.gov/

Barmpatsalou, K., Cruz, T., Monteiro, E., & Simoes, P. (2018). Current and future trends in mobile device forensics. *ACM Computing Surveys (CSUR)*, 51(3). doi:10.1145/3177847.

Chernyshev, M., Zeadally, S., Baig, Z., & Woodward, A. (2017). Mobile forensics: Advances, challenges, and research opportunities. *IEEE Security and Privacy*, 15(6), 42–51. doi:10.1109/MSP.2017.4251107.

DeepSpar Data Recovery Systems: Data Recovery Equipment, Hardware, Business Solutions. (n.d.). Retrieved August 7, 2021, from http://www.deepspar.com/

Digital Evidence Preservation - Digital Forensics - GeeksforGeeks. (2020, June 2). https://www.geeksforgeeks.org/digital-evidence-preservation-digital-forensics/

Dogan, S., & Akbal, E. (2017). Analysis of mobile phones in digital forensics. *2017 40th International Convention on Information and Communication Technology, Electronics and Microelectronics, MIPRO 2017 – Proceedings*, 1241–1244. doi:10.23919/MIPRO.2017.7973613.

EclipseForensics. (2021, March 5). *The Process of Mobile Device Forensics - Eclipse Forensics*. https://eclipseforensics.com/the-process-of-mobile-device-forensics/

Gentry, E., & Soltys, M. (2019). SEAKER: A mobile digital forensics triage device. *Procedia Computer Science*, 159, 1652–1661. doi:10.1016/j.procs.2019.09.335.

Gill, J., Okere, I., Haddad Pajouh, H., & Dehghantanha, A. (2018). Mobile forensics: A bibliometric analysis. *Advances in Information Security*, 70, 297–310. doi:10.1007/978-3-319-73951-9_15.

Guido, M., Buttner, J., & Grover, J. (2016). Rapid differential forensic imaging of mobile devices. *DFRWS 2016 USA - Proceedings of the 16th Annual USA Digital Forensics Research Conference*, 18, S46–S54. doi:10.1016/j.diin.2016.04.012.

Kaushik, K. (2021). A systematic approach to develop an advanced insider attacks detection module. *Journal of Engineering and Applied Sciences*, 8(1), 33. doi:10.5455/JEAS.2021050104.

Kaushik, K., Tanwar, R., & Awasthi, A. K. (2020). Security tools. *Information Security and Optimization*, 181–188. doi:10.1201/9781003045854-13.

Kumar, A., Kuppusamy, K. S., & Aghila, G. (2018). FAMOUS: Forensic analysis of mobile devices using scoring of application permissions. *Future Generation Computer Systems*, *83*, 158–172. doi:10.1016/j.future.2018.02.001.

Maria Jones, G., Godfrey Winster, S., & Scholar, P. G. (2017). Forensics analysis on smart phones using mobile forensics tools. *International Journal of Computational Intelligence Research*, *13*(8), 1859–1869. http://www.ripublication.com

Owen, P., & Thomas, P. (2011). An analysis of digital forensic examinations: Mobile devices versus hard disk drives utilising ACPO & NIST guidelines. *Digital Investigation*, *8*(2), 135–140. doi:10.1016/J.DIIN.2011.03.002.

Sathe, S. C., & Dongre, N. M. (2018). Data acquisition techniques in mobile forensics. *Proceedings of the 2nd International Conference on Inventive Systems and Control, ICISC 2018, 280–286*. doi:10.1109/ICISC.2018.8399079.

Srivastava, A., & Vatsal, P. (2016). Forensic importance of SIM cards as a digital evidence. *Journal of Forensic Research*, *07*(02). doi:10.4172/2157-7145.1000322.

The mobile forensics process: steps and types – Infosec Resources. (2019, July 6). https://resources.infosecinstitute.com/topic/mobile-forensics-process-steps-types/

# 5

## Investigation of Feeding Strategies in Microstrip Patch Antenna for Various Applications

**Kannadhasan Suriyan**

*Cheran College of Engineering, Karur, India*

**Nagarajan Ramalingam**

*Gnanamani College of Technology, Namakkal, India*

## CONTENTS

## 5.1 Introduction

Microstrip antennas are low-profile, lightweight, easy to produce, and suitable with mounting hosts. A microstrip unit consists of two parallel conducting layers separated by a single thin dielectric substratum. The bottom conductor is the ground plane, while the top conductor is a circular/rectangular resonant patch. A number of geometrics may be included in the metallic patch (usually Cu or Au). There are a variety of forms available, including rectangular, circular, elliptical, triangular, helical, ring, and more. Due to its low material cost and simplicity of manufacturing, the microstrip antenna is regarded as the most creative field of antenna engineering that may be carried out inside schools or academic organizations. Prior to the breakthrough of electronic circuit miniaturization and large-scale integration, the idea of a microstrip antenna with a conducting patch on a ground plane separated by a dielectric substrate was underdeveloped. For WiMax deployment with enhanced bandwidth, a stacked tiny slotted antenna was developed. There is a U-port on the bottom patch and a rectangular slot on the top patch. The goal of this project is to reduce the size of a microstrip patch antenna by lengthening the surface current route by cutting a slit in the radiating patch. The antenna seems to be suitable for WiMax applications at frequencies of 3.40–3.69 GHz and 5.25–5.85 GHz. A patch antenna with a microstrip E shape for wireless use. The antenna's polarization may be switched from right-hand circular polarization (RHCP) to left-hand circular polarization (LHCP) and back. At 2.45 GHz, the antenna has a 7% efficient bandwidth and a nominal achieved gain of 8.7 dBi [1–5].

The degree of radiation pattern directivity of an antenna is determined by its gain. That is the radiated power Pr divided by the input power Pi. Due to the conductor and dielectric losses in the products used, a small part of the input power is dissipated and converted into radiated and surface wave power. Instead of simply the radiation efficiency, the total utility may be used to define antenna gain. The high gain antenna offers the benefit of greater range and signal strength, but it must be directed precisely in certain directions. A mathematical function or graphical representation of the antenna's radiation properties as a function of spatial coordinates is known as the radiation pattern. This is the proportion of an antenna's total power output to its input power. VSWR = Vmax/Vmin is the formula for calculating the voltage standing wave ratio. It should be between 1 and 2 on the scale. In a standing wave pattern, Voltage Standing Wave Ratio (VSWR) is defined as the ratio of highest to minimum voltage. When energy was reflected off a load, a standing wave was created. Improper impedance matching is to blame.

## 5.2 Microstrip Patch Antennas

The microstrip patch antenna has a variety of advantages over conventional microwave antennas. The microstrip patch antenna is a popular subject in antenna theory because of its many advantages over conventional antennas, including low cost, lightweight, simplicity of feeding, and appealing radiation properties. Since the analysis must concentrate on lowering antenna size while retaining acceptable gain and bandwidth, researchers have been interested in this subject. This article covers antenna array topologies, composite antennas, highly integrated antenna/array and feeding network, operating at relatively high frequencies, sophisticated manufacturing methods, and various approaches suggested to enhance microstrip patch antenna performance during the past several decades.

The restricted bandwidth of these antenna designs is one of their main drawbacks. Increased substrate thickness is the fastest way to improve bandwidth, but it has a major drawback: it lowers output since a large portion of the input power is lost in the capacitor, reducing the available power that the antenna can radiate. Reduced structural height, on the other hand, may seem to be a cost-effective alternative, but it may lead to worse bandwidth impedance and radiation efficiency. This is also a tradeoff in the construction of lightweight antennas that are nonetheless functional. As a consequence, several additional enhanced methods, such as the use of high permittivity dielectric resonators for microstrip patch antennas, are utilized to offer wide-impedance bandwidths for microstrip antennas. Microstrip antenna bandwidth is optimal for low dielectric constant substrates as a proportion of total avalanche bandwidth, thus high permittivity substrates are a bad option

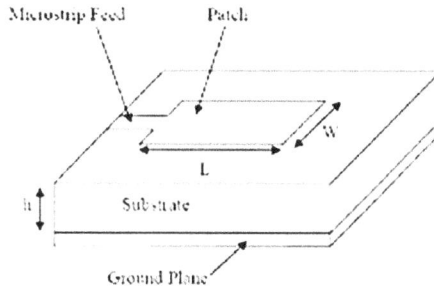

**FIGURE 5.1**
Microstrip patch antenna.

for antenna bandwidth. Depending on the procedure, these methods may involve both touch and non-contact methods. During the contacting phase, radio frequency (RF) power is sent directly to the radiating patch through a connecting component such as a microstrip thread. In the non-contacting phase, electromagnetic coupling transmits power between the microstrip line and the radiating patch. The microstrip thread, coaxial probe (both contact systems), aperture coupling, and proximity coupling are the four most prevalent feeding techniques (both non-contacting schemes).

In recent years, the microstrip antenna field has seen a lot of creative work, and it is now one of the most active fields in the communication industry. Cellular phones, broadband connection, wireless local area networks (WLANs), and cellular phone infrastructure are among the fastest-growing economic sectors today. The microstrip antenna is a popular option because to its tiny weight, low volume, low manufacturing cost, and dual and triple frequency operating capabilities. On the other hand, microstrip antennas have a number of disadvantages. The limited bandwidth of microstrip patch antennas is a significant drawback. Various ways for enhancing it have been proposed, cutting holes in it being one of the methods that has been supported by many investigations. In our experiment, we cut two parallel slots in a rectangular microstrip antenna to mimic an e-shaped patch antenna operating in the X-band at 8.6 GHz. The suggested antenna is shown in Figure.5.1 placed on a Rogers-Corp 0.1575 cm RT DUROID 5880 substrate with a dielectric constant of 2.2 and a tangent loss of 0.00044 and a dielectric constant of 2.2 [6–10].

## 5.3 Electromagnetic Bandgap Structure

The usage of electromagnetic band gap (EBG) systems in microwave culture has gotten a lot of interest in recent years due to its unique features. In addition to traditional leading and/or filtering structures, these structures are intermittent in nature and prohibit all electromagnetic surface waves from transmitting within a specific frequency range, known as the band distance, allowing for more control of electromagnetic wave activity. Dummy EBG patterns, which are etched on the feed line that connects the two patches, are lightweight and tiny in size, and they provide a substantial bandwidth gain (48.8% increase with a bandwidth). When utilizing an impedance matching circuit for impedance matching, the radiating patch's radiation properties should not be altered. Furthermore, the Impedance Matching (IM) method may be used for any patch substrate. Similarly, slot

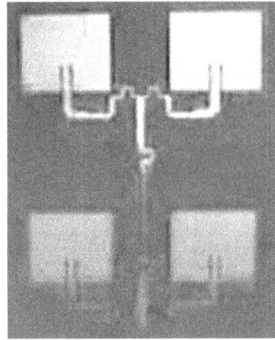

**FIGURE 5.2**
Electromagnetic band gap structure.

antennas are unidirectional microwave antennas that create slots in microstrip antennas by utilizing the ground plane's L, U, T, and inverted T slots to produce Ultra-Wideband signals (UWB).

A single approximately rectangular microstrip radiator reactively filled with active negative capacitor and composite-resonator microstrip antennas using metamaterial resonators with broad bandwidth and high gain may be used to add negative capacitor/inductor high gain. The poor gain of traditional microstrip antenna components is another issue that must be addressed. The phrase optimization of topology is frequently used to describe the most common category of configuration optimization methods in which the forms, as well as the communication of particular system components, are subject to design. Array topologies, such as dual patch antenna arrays, are an example of this. The cavity backing technique has been used to remove the bidirectional radiation pattern, which is used to have greater gain relative to traditional microstrip antennas, using the material distribution method, in which the geometry domain is divided into small elements that together reflect a picture of the system, the most common form of carrying out topology optimization. Lens shielding is a different way to boost your benefits. Canonical lenses like hemi elliptical, elliptical, hyper-hemispherical, and extended hemispherical lenses focus radiation from radiator components. By integrating microstrip radiator components with dielectric lenses, the combined microstrip lens antenna may be considered a hybrid antenna that is very effective for applications of high frequencies (mm, sub-mm, terahertz (THz), and optical waves; see Figure 5.2). It's also common knowledge that using an antenna array to boost gain is a good idea [11–15].

## 5.4 Metamaterial Structure

Metamaterial technologies may be used in conjunction with patch antennas to enhance output criteria. This research focuses on the metamaterial construction of a high gain circular waveguide array antenna. The metamaterial is made out of a square lattice of copper grids. The usage of metamaterial composition increases the electric field as the electromagnetic wave propagates in free space. A metamaterial construction's antenna gain improves from 9053 to 17.34 dB. The metamaterial framework's gain for the circular waveguide aperture antenna is now extremely near to the theoretical optimum antenna value for the same size and operating frequency. The array layout, when coupled with metamaterial-mantled technology, is a more potent way to maximize advantage. When compared to a conventional antenna array, simulation results indicate a gain of around 7 dB in the antenna array, indicating that the metamaterial system of the antenna array's radiation characteristics is substantially improved. Microstrip patch antennas are further miniaturized using metamaterials. For C band applications, patch antennas made of metamaterials may be used. The area of such an antenna is decreased by a factor of 2.4, and the gain directivity rises from 4.17 to 5.66 dBi in metamaterial architecture, compared to 4.17 dBi in conventional design. The metamaterial substrate may be built in a variety of ways to function at various frequencies. When rendering metamaterial antenna substrates, framed square rings, various C patterns, square and circular patterns, and other designs are taken into account. All of these solutions are intended to increase bandwidth and reduce return loss while reducing height. An antenna's permeability and permittivity may be investigated in a number of methods. The wave interference system, the Nicolson Ross Wier approach, the National Institute of Standards and Technology (NIST) iterative methodology, current pre-iterative technology, and the short circuits technique are all examples of these techniques. The Nicolson–Ross–Weir (NRW) technique was used to estimate the permittivity and permeability of the proposed frameworks in most experiments.

A circuit design with a wider range of material parameters resulting in a negative refractive index has yielded revolutionary antennas. Contemporary antenna layouts surpass traditional antenna designs by combining a left-handed transmission line segment with a normal (right-handed) transmission line. The left-handed transmission lines are basically a phase-advance filter with a high pass. Right-handed transmission lines, on the other hand, act as a step-lag low-pass filter. The composite right/left-handed (CRLH) metamaterial will be used in this arrangement. One of the most frequent applications of metamaterials is in the construction of wide-band antennas. In the metamaterial antenna design, the hybrid right-/left-handed transmission line approach is used to enhance the antenna's efficiency. A Mushroom Structured Composite Right-/Left-Handed transmission line (CRLH-TL) metamaterial may shrink by 61.11% in size. Furthermore, a broad band may be obtained by lowering the ground plane of the antenna. A lightweight ultra-wide band (UWB) antenna may be made using a metamaterial framework. The antenna has an 189% broad bandwidth. A single patch antenna's bandwidth may be increased by carefully arranging a number of metamaterial unit cells [16–19].

Because of their fascinating and unexpected properties, metamaterials have played a major role in antenna design. Metamaterial-based antennas may enhance the efficiency of microstrip antennas, according to the findings of this research. By reactively loading the metamaterial framework above the substrate, a metamaterial antenna is produced. Metamaterial substrates are available in many different forms and sizes. If the metamaterial substrate is modified in any manner, the antenna parameters will improve. A wideband antenna may be created by merging a number of metamaterial unit cells. When metamaterial technologies are used, the gain of a patch antenna rises by 1.5–7 dB. Metamaterials are characterized by their miniaturization. In all of the experiments described here, the introduction of metamaterials results in a patch antenna size decrease of about 50%. The microstrip patch antenna has two main flaws: low gain and restricted bandwidth. Figure 5.3 shows a metamaterial antenna that successfully addresses the above-mentioned microstrip patch antenna problems while also allowing for miniaturization [20–25].

**FIGURE 5.3**
Metamaterial structure.

## 5.5 Multiple Input and Multiple Output (MIMO) Systems

Due of its tremendous spectrum performance, MIMO systems employing wideband microstrip patch antennas are the fastest developing area of technology that has sparked the interest of daily life in future wireless communications. During the next decade, the most urgent problems in wireless communications will continue to be spectral quality and congestion. The range would be pushed to its limits by cellular Internet, as well as smartphone video and data transmission. Increased bandwidth usage would result in increased network interference. Given this, the most essential objective is to improve spectral performance while also including steps to minimize disturbance. MIMO systems have a broad range of data rates and high spectral efficiency. The importance of microstrip antennas and the many types of microstrip antennas used in the construction of MIMO systems are explored in the current wireless scenario. These antennas are narrowband components whose use in today's high-data-rate wireless networks is limited. Another significant problem that affects the effectiveness of MIMO systems is reciprocal coupling, which arises due to the decreased distance between the antennas. This article analyses the microstrip patch antenna's technical developments over the past 40 years. A lot of research is being done on microstrip antennas in order to make the most use of them in the next generation of wireless contact ions. MIMO is a well-known method for making future wireless broadband data networks more efficient. As a consequence of adjusting for the microstrip antenna's strength and bandwidth, more solutions are developing. According to the survey, different possibilities of new wideband microstrip patch antenna to boost the features of MIMO can be discovered by using techniques such as adding parasitic factor in a coplanar or stack configuration, increasing substrate thickness, changing the shape of a patch, and inserting slots. Following that, the survey reveals that impedance and microstrip patch antenna properties may be improved.

Technology that was unimaginable just a few decades ago is now possible because to fast advancements in information technology and wireless networking. Our lives have traditionally revolved around personal contact. Almost everything is now cellular and portable. Other components should be maintained as far away from their surroundings as feasible, whereas an antenna should effectively radiate to free space in a predictable manner. An increasing variety of wireless connection protocols are being used in systems due to the enhanced capabilities of smartphones. Even if they are currently in use, devices must be able to manage an ever-increasing number of antennas. A substantial increase in bandwidth will be required. Meanwhile, the decreasing unit size has resulted in a vastly increased space restriction in antenna deployment situations. Microstrip Antennas (MSAs)

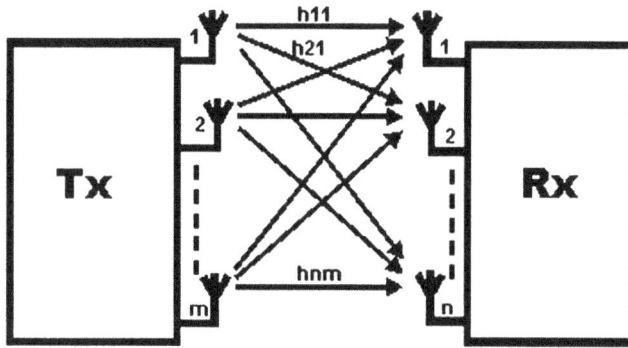

**FIGURE 5.4**
MIMO antenna.

are also constructed using printed-circuit technology to achieve high mass manufacturing at a low cost. MSAs, which are used for military and industrial purposes, are replacing many traditional antennas, as seen in Figure 5.4.

## 5.6 Techniques of Microstrip Patch Antennas

Microstrip is an electrical transmission line constructed of Printed Circuit Board (PCB) that is used to carry microwave frequency communications. It has a strip that serves as a substratum and is separated from the ground plane by a dielectric layer. A Microstrip Patch Antenna (MPA) is made up of a conducting patch on one side and a ground plane on the other of a planar or non-planar dielectric substrate. For narrowband microwave wireless communications that need semi-hemispheric coverage, it is a common written resonant antenna. The MPA has gotten a lot of attention because of its planar design and simplicity of integration with microstrip technology, and it's often used as an array element. A wide range of patch antennas for microstrips has been explored to date. A large list of geometries and their unique characteristics may be found. The most popular and frequently used microstrip antennas are rectangular and circular patches. Both the simplest and the most complex coders utilize these fixes. The most popular and frequently used microstrip antennas are rectangular and circular patches. Both the simplest and the most complex coders utilize these fixes. These antennas have a low profile, are compatible with planar and non-planar surfaces, are quick and inexpensive to manufacture using state-of-the-art printed-circuit technology, are mechanically stable when installed on rigid surfaces, are compatible with Monolithic Microwave Integrated Circuit (MMIC) designs, and have a wide range of patch shapes and modes.

It's important to differentiate between communicative and non-contacting methods. RF power is sent directly to the radiating patch through a connecting device such as a microstrip thread during the contacting process. Electromagnetic field coupling is utilized to transmit power between the microstrip line and the radiating patch in the non-contacting technique. Microstrip thread, coaxial probe (both contacting systems), aperture coupling, and proximity coupling (both non-contacting schemes) are the four most common feeding techniques (both non-contacting schemes).

The antenna efficiency factor of MPAs is very high (Q). It shows antenna-based losses, in which a large Q limits bandwidth and results in poor performance. To decrease Q, the dielectric substrate thickness may be raised. However, as the thickness of the layer rises, a greater percentage of the source's total power is transmitted through surface waves. It's essential to see this contribution from surface waves as excessive. Antenna characteristics degrade when power is lost as it passes through the dielectric bends. Other issues, such as reduced gain and power handling capabilities, may be solved by arranging the component in an array.

MPAs may be fed in a number of ways. Contact and non-contact methods are examples of such strategies. RF power is sent directly to the radiating patch through a connecting device such as a microstrip thread during the contacting process. In this technique, a conducting strip is attached to the tip of the microstrip patch precisely. The conducting strip has a smaller diameter than the patch. This kind of feed arrangement also has the advantage of enabling the feed to be etched on the same substrate as the rest of the board, allowing for a flat layout. In this arrangement, the coaxial connection's inner conductor extends across the dielectric and is connected to the radiating patch, while the ground plane is attached to the outer conductor. The feed may be inserted within the patch at any of the 26 places needed to balance its input impedance, which is the most important benefit. It has the drawbacks of low bandwidth and being difficult to model. In this technique, the ground plane separates the radiating patch from the microstrip feed line. The patch is connected to the feed line in the ground plane via a slot. The coupling slot underneath the patch is oriented due to configuration symmetry, resulting in little cross polarization. Spurious radiation is reduced because the ground plane separates the patch and feed sides. The most major drawback of this feed technique is that it is expensive to produce owing to the many layers; it also adds to the antenna's thickness. This kind of feeding method is referred to as an electromagnetic connection system. There are two dielectric substrates with a feed line going between them. On top of the higher substratum, the radiating patch rests. The primary benefit of this feed technique is that it avoids stray feed radiation while still providing very high bandwidth (up to 13%). Due to the two dielectric layers that need perfect synchronization, the major drawback of this feed method is that it is difficult to produce.

## 5.7 Advantages and Disadvantages

Because of its low-profile form, MPAs are becoming increasingly popular in wireless applications. As a consequence, they function well with cellular mobile devices like phones and pagers that have inbuilt antennas. On warheads, telemetry and contact antennas must be narrow and conformal, and are often MPAs. Satellite networking is another area where they've been effective.

The following are the main advantages of MSAs:

1. It is light in weight and holds a tiny amount of information.

2. A low-profile planar arrangement that can be readily transferred to the host's surface.

3. Low-cost production that can be done in big batches.

4. In both linear and circular directions, polarization is assisted.

5. Microwave integrated circuits (MICs) are simple to combine with other circuits.

6. Operation on several frequencies is feasible.

7. Mechanically stable on stiff surfaces when built.

The following are the many disadvantages:

1. A little advantage.

2. Lack of efficacy.

3. Higher Ohmic losses.

4. Feeding structure that is dynamic.

5. Polarization purity is difficult to maintain.

6. The amount of power that can be handled is restricted.

## 5.8 Applications

Because of their reliability and sturdy construction, MPAs are popular. MPAs are used in a wide range of applications, including medical, satellite, and military applications such as rockets and aircraft weaponry. Because of the low cost of substrate composition and production, they are presently thriving commercially. The MPA may be used for a variety of purposes. Because of their apparent and visible advantages, microstrip antennas have a broad variety of applications, although they do have certain downsides, such as low gain, restricted bandwidth, and larger size. This article describes a few additional improved regions in relation to these drawbacks based on previous work on microstrip antennas. These drawbacks may be mitigated by utilizing a range of microstrip antenna topologies, microstrip antenna-based composite antenna construction, and improved manufacturing methods for microstrip antennas to securely incorporate higher frequency antennas while simultaneously juggling these three variables. This page provides an overview of microstrip antennas as well as current MPA development for a variety of applications. Over the past 15 years, microstrip antenna technology has been recognized as an emerging trend in the antenna business. A microstrip antenna's bandwidth is known to be limited, and it is believed to be proportionate to the patch size. The patch grows broader and more voluminous when the patch size is raised to enhance bandwidth. To satisfy the bandwidth restriction, a multi-layer patch was developed to reduce the efficient dielectric constant without increasing the size significantly. This means that, although the size is essentially same, the bandwidth is much increased. Quality parameters including bandwidth and reflection loss have been assessed and compared to single layer dielectric substrates.

The MPA satisfies all of the criteria, and many microstrip antennas have been developed for use in mobile communication systems. Circularly polarized radiation patterns are needed for satellite communication, which may be produced using either square or circle patches of one or two feed points. Microstrip patch antennas with sintered substrate material and high permittivity for the global positioning system (GPS). There are very tiny circularly polarized antennas. Mobile networking, distribution, engineering, transportation, and health care are just a few of the sectors that use radiofrequency identification (RFID). Depending on the application, RFID devices typically operate at frequencies ranging from 30 Hz to 5.8 GHz. A tag or transponder and a transceiver or reader make up an RFID system. The IEEE 802.16 standard is referred known as WiMax. It should have a theoretical

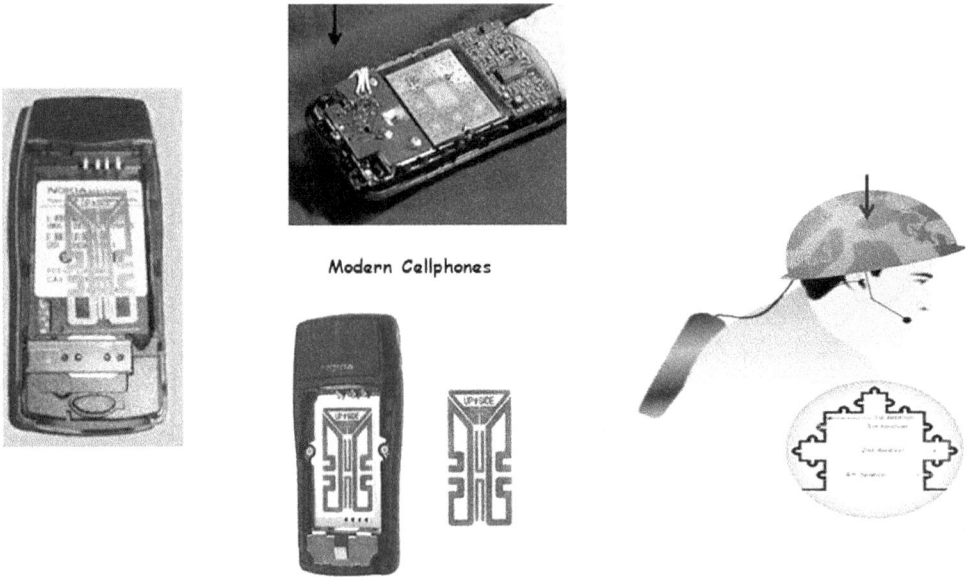

**FIGURE 5.5**
Applications of microstrip patch antenna.

range of 30 miles and a data rate of 70 megabits per second. The Microstrip patch antenna may be used in WiMax contact devices and produces three resonant modes at 2.7, 3.3, and 5.3 GHz. Radar may be used to track moving objects like people and cars. Microstrip antennas are a fantastic choice. Photolithography-based manufacturing technique allows for the mass manufacture of microstrip antennas with repeatable output at a lower cost and in less time than traditional antennas. The microstrip antenna in this instance works in the Industrial, Scientific, and Medical (ISM) band of 2400–2484 MHz. The microstrip antenna, despite the inclusion of an air layer, only occupies a limited area of 33.3-6.6-0.8, as shown in Figure 5.5.

## 5.9  Tools of Microstrip Patch Antenna

High-Frequency Structure Simulation (HFSS) is a commonly used 3D full-wave electromagnetic field modeling tool in the industry. HFSS calculates E- and H-fields, waves, S-parameters, and radiated field effects near and far. Because of its automated solution method, HFSS is successful as an engineering design technique when users are simply allowed to provide geometry, material properties, and necessary performance. HFSS can then create a suitable, successful, and precise mesh to solve the problem automatically. The IE3D electromagnetic modeling and optimization program from Zeland Software Inc. is useful for circuit and antenna design. To handle current distribution on general form 3D and multi-layer structures, IE3D employs a full-wave, method-of-moments field solver, as well as a menu-driven visual interface with automated meshing for model building.

Advanced Design Framework is a leading electrical design automation software for RF, microwave, and high-speed wireless systems in the globe. Advanced Design Simulator

(ADS) is used by leading businesses in the wireless connection, networking, aerospace, and military sectors to pioneer the most creative and commercially popular technologies, including X-parameters and 3D EM simulators, in an efficient and user-friendly interface. ADS offers complete, standard-based design and verification for WiMAX, LTE, multi-gigabit per second data connections, radar, and satellite applications using Wireless Libraries and circuit-system-EM co-simulation on an embedded platform.

The Computer Simulation Technology (CST) Microwave Studio (CST MWS) is a specialized platform for high-frequency 3D EM modeling. Due to its outstanding performance, CST MWS has become the preferred choice in large technical R&D departments. CST MWS is used to evaluate High-Frequency (HF) instruments such as antennas, filters, couplers, planar, and multi-layer structures, as well as Scientific Identification (SI) and Electromagnetic Compatibility (EMC) effects.

## 5.10 Parameters of Microstrip Patch Antenna

An antenna is a crucial part of any wireless communication system. An efficient antenna design reduces device demands while improving overall efficiency. A number of variables referred to as antenna characteristics, influence the antenna's effectiveness: The reflection of signal strength in a transmission line as a consequence of a system's installation is known as return loss. As a consequence, the Return Loss (RL), like the VSWR, is a measure of how well the transmitter and antenna are matched.

$$\text{The RL is calculated as follows: } RL = -20 \log 10 \, (\text{about}) \, dB \, dB$$

The optimum power transfer between the transmission line and the antenna is typically determined by input impedance. This transition happens when the antenna input impedance and the transmission line input impedance are both balanced. If the waves do not fit, they will be reflected and returned to the energy source at the antenna termination. The overall performance of the gadget suffers as a consequence of this representation of energy outcomes. It's also critical for the antenna's input impedance to be largely resistive since this implies that a large portion of the power provided to it will be radiated. The input impedance is made up of real and complex components, and it has the following basic shape:

$$Rin + jXin = Zin$$

Rin denotes the impedance radiating component's resistance or strength.

Xin denotes the impedance's reactive portion, also known as the power storage section.

The voltage standing wave ratio is the ratio between a transmission line's highest and lowest voltages (VSWR). The VSWR shows how well the antenna terminal's input impedance matches the transmission line's characteristic impedance, which may be calculated using the levels of reflected and incident waves. As the VSWR rises, the mismatch between the transmission line and the antenna grows. A reduction in VSWR indicates that a good match has been obtained for a minimal VSWR. The impedance of many wireless networks is 50 Ohm. The antenna's impedance must also be as near to 50 ohms as possible. An antenna impedance of precisely 50 ohms implies a VSWR of 1. A VSWR ratio of 1.5:1 is needed for the feasible antenna.

The antenna gain is a measure of how effective the antennas are overall. An antenna with a gain equal to its directivity is said to be 100% efficient. The average performance of an antenna is influenced by a number of variables. An antenna's radiation patterns include descriptions of how the antenna focuses the energy it produces. If all antennas are 100% effective, they will emit the same total energy for identical input capacity, regardless of the pattern form. The majority of radiation patterns are observed on a relative power scale. It may be seen in a 360-degree polar plot. The radiation pattern is shown in Figure 5.1. When presenting antenna pattern results, the convention of an E-plane and H-plane pattern is used in many cases. The antenna's emitted electrical field potential is contained in the E-plane, while the antenna's radiated magnetic field potential is contained in the H-plane.

D is an integral value that indicates how effectively an antenna can focus emitted energy. The ratio of total radiated to radiated reference antenna is known as directivity. In most cases, the reference antenna is an isotropic radiator with one and the same radiated radiation directionality in both dimensions.

The polarization of an antenna determines the direction and type of the electric field vector of radiated waves. Polarization may be divided into three categories:

1. Polarization along a straight line.
2. Polarization on an elliptical axis.
3. The circular polarization is the third step in the polarization process.

Most antennas transmit in either linear or circular polarization. Antennas with linear polarization emitted in the same plane as the wave propagation direction, whereas antennas with circular polarization emitted in a circular pattern.

The frequency spectrum in which an antenna may satisfy the output requirements of a set of standards is referred to as bandwidth. The performance trade-offs between all of the performance parameters mentioned above are the most important thing to grasp about bandwidth.

The difference between the gain from the front of the antenna and the gain from the back of the antenna is the front to rear isolation ratio. If the antenna is used in a crowded frequency band, communication engineers are interested in the Front to Back Ratio (FBR). When comparing Yagi-Uda antennas, amateur radio operators often use front to back isolation as a criterion.

Any quantitative aspects of the antenna pattern features may be stated until the antenna pattern information is given in a polar map. The 3 dB beamwidth (1/2 power stage), directivity, side lobe level, and front to back ratio are all quantitative properties. Start with the most basic reference antenna, the point source, to better grasp these principles. A hypothetical antenna called a point source distributes energy equally in both directions, resulting in a perfect antenna pattern. These antennas have a directivity of 0 dB and are described as an omnidirectional isotropic radiator. In actuality, since the antenna is described as omnidirectional, it is believed that this only applies to the horizontal or azimuth sweep plane. The angular breadth of the −3 dB points in the antenna's pattern is simply compared to the maximum pattern to get the antenna's 3 dB beam width. The pattern's −3 dB points correspond to the places where the power level is 3 dB lower than the pattern's highest value. In general, each of the numerous pattern sweep plane antennas' 3 dB beamwidth is specified separately; there will always be some unique path of maximum radiated energy. The various journals that have been published on microstrip patch antenna are shown in Table 5.1 and Figure 5.6.

**TABLE 5.1**

Number of Paper Published in Peer-Reviewed Journal

| Year | Number of Papers Published | | | | |
|------|------|------|---------|-------|------|
|      | IEEE | IET  | Hindawi | Wiley | ACM  |
| 2000 | 10   | 5    | 10      | 20    | 10   |
| 2001 | 15   | 10   | 15      | 30    | 15   |
| 2002 | 12   | 15   | 20      | 32    | 20   |
| 2003 | 16   | 22   | 25      | 35    | 25   |
| 2004 | 18   | 28   | 30      | 38    | 30   |
| 2005 | 20   | 30   | 35      | 40    | 35   |
| 2006 | 22   | 35   | 20      | 45    | 20   |
| 2007 | 25   | 38   | 25      | 48    | 25   |
| 2008 | 30   | 40   | 28      | 52    | 40   |
| 2009 | 40   | 45   | 25      | 55    | 45   |
| 2010 | 45   | 48   | 30      | 59    | 48   |
| 2011 | 48   | 52   | 35      | 62    | 52   |
| 2012 | 50   | 56   | 40      | 65    | 56   |
| 2013 | 55   | 52   | 42      | 70    | 52   |
| 2014 | 65   | 58   | 45      | 80    | 48   |
| 2015 | 72   | 60   | 48      | 85    | 49   |
| 2016 | 85   | 62   | 49      | 86    | 52   |
| 2017 | 55   | 65   | 52      | 84    | 57   |
| 2018 | 80   | 68   | 57      | 86    | 60   |
| 2019 | 78   | 70   | 60      | 88    | 62   |
| 2020 | 70   | 70   | 62      | 90    | 48   |

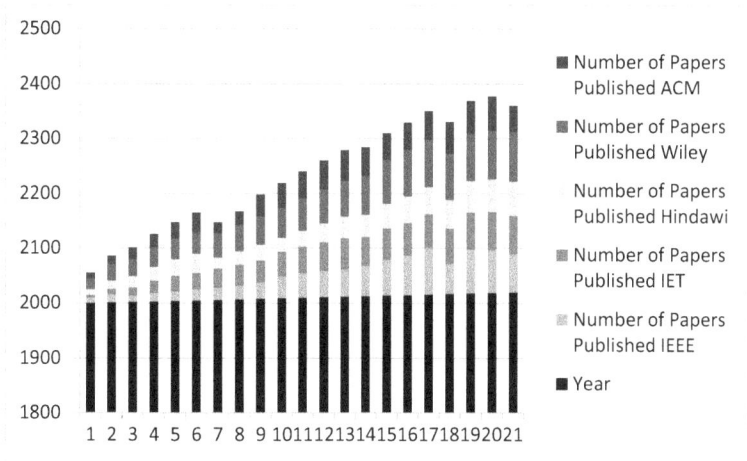

**FIGURE 5.6**
Comparison chart for various journals.

## 5.11 Conclusion

This article analyses the MPA's technical developments over the past 40 years. An antenna is a kind of transducer that transforms RF signals into alternating current (AC) or the other way around. Radio signals are sent and received using both receiving and transmitting antennas. Antennas are necessary for the proper functioning of any radio equipment. The microstrip antenna is becoming more technologically advanced every day. The microstrip antenna is undergoing extensive study in order to improve its use in the future. As a consequence of adjusting for the microstrip antenna's strength and bandwidth, more solutions are developing. According to the study, few publications on the microstrip antenna were written in the early years, but there was a rise in numbers until 2003, when the number of published papers began to decline. Medical, military, mobile, and satellite communications are just a few of the applications for microstrip antennas. Following this review of the literature, it is thought that the circular microstrip antenna may be built for various frequency bands and that different feeding methods with different substratum content might be utilized. In comparison to other methods, strip line feeding is a simple approach. This article provides an overview of microstrip patch antennas. A slotted fix will improve weak gain and limited power handling capabilities, according to various research reports. In this study, several feeding techniques and fundamental antenna settings were examined.

## References

1. Taflove A, Hagness SC (2005). *Computational Electrodynamics, the Finite-Difference Time-Domain Method*. 3rd ed. Artech House: Norwood, MA.
2. Zheng F, Chen Z, Zhang J (2000). Toward the development of a threedimensional unconditionally stable finite-difference. *IEEE Trans Microw Theory Tech* 48(9):1550–1558.
3. Zheng F, Chen Z (2001). Numerical dispersion analysis of the unconditionally stable 3-D ADI-FDTD method. *IEEE Trans. Microw. Theory Tech.* 49(5):1006–1009.
4. Huang BK, Wang G, Jiang YS, Wang WB (2003). A hybrid implicitexplicit FDTD scheme with weakly conditional stability. *Microw Opt Technol Lett.* 39(2):97–101.
5. Chen J, Wang J (2007). A three-dimensional semi-implicit FDTD scheme for calculation of shielding effectiveness of enclosure with thin slots. *IEEE Trans Electromagn Compat.* 49(2):354–360.
6. He XJ, Wang Y, Wang JM, Gui TL, Wu Q (2011). Dual-band terahertz metamaterial absorber with polarization insensitivity and wide angle. *Prog Electromagn Res.* 115:381–397.
7. Kumari K, Mishra N, Chaudhary RK (2017). An ultra thin compact polarization insensitive dual band absorber based on metamaterial for X-band applications. *Microw Opt. Technol. Lett.* 59(10):2664–2669.
8. Wang GD, Chen JF, Hu XW, Chen ZQ, Liu M (2014). Polarizationinsensitive triple-band microwave metamaterial absorber based on rotated square rings. *Prog Electromagn Res.* 145:175–183.
9. Ghosh S, Member S, Bhattacharyya S (2015). An ultra wideband ultrathin metamaterial absorber based on circular Split rings. *IEEE Antennas Wirel Propag Lett.* 14:1172–1175.
10. Oraizi H, Valizade Shahmirzadi N (2017). Frequency- and time-domain analysis of a novel UWB reconfigurable microstrip slot antenna with switchable notched bands. *IET Microwave Antennas Propag.* 11(8):1127–1132.
11. Tuffy N, Guan L, Zhu A, Brazil TJ (2012). A simplified broadband design methodology for linearized high-efficiency continuous class-F power amplifiers. *IEEE Trans.Microw Theory Techn.* 60(6):1952–1963.

12. Sun YJ, Zhu XW (2015). Broadband continuous class-F−1 amplifier with modified harmonic-controlled network for advanced long term evolution application. *IEEE Microw Wireless Compon Lett*. 25(4):250–252.

13. Matthaei GL (1964). Tables of Chebyshev impedance-transformation networks of low-pass filter form. *Proc IEEE*. 52(8):939–963.

14. Giofre R, Colantonio P, Giannini F, Piazzon L (2007). A new design strategy for multi frequencies passive matching networks. Paper presented at: *Proceedings of the European Microwave Conference*, London. 838–841.

15. Tasker PJ, Benedikt J (2011). Waveform inspired models and the harmonic balance emulator. *IEEE Microw Mag*. 12(2):38–54.

16. Taga T (1990). Analysis for mean effective gain of mobile antennas in land mobile radio environments. *IEEE Trans Veh Technol*. 39(2):117–131.

17. Morsy MM, Morsy AM (2018. Dual-band meander-line MIMO antenna with high diversity for LTE/UMTS router. *IET Microw Antennas Propag*. 12(3):395–399.

18. Choukiker YK, Sharma SK, Behera SK (2014). Hybrid fractal shape planer monopole antenna covering multiband wireless communications with MIMO implementation for handheld mobile devices. *IEEE Trans Antennas Propag*. 62(3):1483–1488.

19. Manteghi M, Rahmat SY (2015. Multiport characteristics of a wide-band cavity backed annular patch antenna for multipolarization operations. *IEEE Trans Antennas Propag*. 53(1):466–474.

20. Bhat B, Koul SK 1989. Stripline-like transmission lines for microwave integrated circuits. *New Age Int*. https://bruceviteun.files.wordpress.com/2016/11/631_full.pdf.

21. Alhegazi A, Zakaria Z, Shairi NA, Ibrahim IM, Ahmed S (2017). A novel reconfigurable UWB filtering-antenna with dual sharp band notches using double Split ring resonators. *Prog Electromagnet Res C*. 79:185–198.

22. https://www.macom.com/products/product-detail/MA4SPS552

23. Zhang C, Gong J, Li Y, Wang Y (2018). ZerothOrder-mode circular microstrip antenna with patch-like radiation pattern. *IEEE Antennas Wireless Propag Lett*. 17(3):446–449.

24. Li Y, Li W, Ye Q (2013). A reconfigurable triple-notch-band antenna integrated with defected microstrip structure band-stop filter for ultra-wideband cognitive radio applications. *Int J Antennas Propag*. 2013:472645.

25. Amir H, Nazeri A, Falahati RME (2019). A novel compact fractal UWB antenna with triple reconfigurable notch reject bands applications. *Int J Electron Commun*.;101:1.

# 6

## Optical Encryption of Images Using Partial Pivoting Lower Upper Decomposition Based on Two Structured Deterministic Phase Masks in the Hybrid Transform Domain

**Priyanka Maan and Hukum Singh**

*The NorthCap University, Gurugram, India*

**A. Charan Kumari**

*Dayalbagh Educational Institute, Agra, India*

## CONTENTS

## 6.1 Introduction

With the fast advancement of data innovation, information technology and the expanding use of the internet, computerized items are generally connected in every aspect of an individual's life. Despite the fact that these items convey comfort to individuals, they additionally bring an ever-increasing number of major issues in the meantime, for example, data encroachment and unauthorized altering. As one of the vital approaches to stay away from infringement and unauthorized updates, image encryption is getting increasingly more

DOI: 10.1201/9781003206088-6

attention. Owing to the speed, multi-dimensional nature, parallel processing capability, the optical methods utilized in data security frameworks shows an undeniably significant job and has drawn substantially more consideration. The first ones who proposed a spearheading plan named double random phase encoding (DRPE) in 1995 to secure information using optical technique were Refregier and Javidi [1]. Thereafter, different schemes utilizing DRPE strategy [2–7] have been broadly utilized in data security frameworks. Along these lines, formal optical cryptanalysis has likewise been done to assess the security level of optical cryptosystems. It has been discovered that due to their intrinsic linearity, DRPE-based cryptosystems are weak against different attacks [8–18]. To address the issue, different optical algorithms have been proposed. Different techniques including phase retrieval [19–22], compressive sensing [23–26], spherical wave illumination [27] are additionally discovered and utilized to achieve secure image encryption plans. But every algorithm has its own drawback like phase retrieval technique use iterative procedures which are complex and difficult to achieve optically.

The cryptosystems based on DRPE are prone to various attacks, for example, known-plaintext attack (KPA), chosen plaintext attack (CPA), cipher text only attack (COA) etc. which shows that schemes based on DRPE are unreliable in view of their symmetric nature. To remove this drawback, researchers developed cryptosystems that are asymmetric in nature that implies encryption and decryption keys are different. The first asymmetric optical encryption technique called phase-truncation Fourier transform was developed by Qin and Peng [28] that removes the linearity problem of DRPE and is resistant to basic attacks. Following this, they presented [29] the properties of the two decryption keys used in PTFT called universal and special key. Later on, this phase-truncation strategy has been extended to different transforms like Fresnel transform [30], Gyrator transform [31], fractional Fourier transform [32] etc. Wang and Zhao [33] implemented a unique attack on PTFT-based cryptosystem and likewise proposed an improvement in PTFT-based cryptosystem to oppose this particular attack [34]. Various other asymmetric cryptosystems are developed for images with significant techniques operating on amplitude and phase-truncated Fresnel transform [35], using discrete Haar-wavelet transform for image hiding [36]. The main idea behind improved PTFT-based asymmetric cryptosystems is to increase the number and complexity of private key. Following it, a cryptosystem [37] using PTFT and Joint transform correlator (JTC) has been designed to increase the position set parameter. Afterwards, the change in the structure of public and private keys and some unique feature addition to the cryptosystem is treated as a primary concern in design of asymmetric cryptosystem. Following this many researchers developed various masks that act as keys in symmetric and asymmetric cryptosystems. Some work on radial Hilbert mask [38,39], devil vortex phase structure [40], combination of Kronecker product and hybrid phase mask [41], deterministic phase mask [42], structured phase mask [43,44], chaotic and logistic maps [45,46], and rear mounted masks [47]. Every cryptosystem has its own pros and cons [48]. Then the cryptosystems using PTFT with random amplitude mask (RAM) [49], PTFT and interference with random phase masks (RPM) [50] were developed, but later it was found that they can be attacked [51,52]. Subsequently, the cryptosystems were developed based on different hybrid transforms [53].

To increase the security level further, researchers started using biometric keys for image encryption [54], chaotic fingerprint phase mask based color image cryptosystem [55], 3D vector decomposition technique [56], multiple image cryptosystem using equal modulus decomposition and frequency modulation [57], phase retrieval technique for image encryption and authentication using digital signature [58], color image based compact cryptosystem using unequal modulus decomposition [59], optical encryption and authentication scheme based on phase-shifting interferometry in a JTC [60], optical cryptosystem using lower upper (LU) decomposition in light of Gyrator transform domain [61]. Further, the cryptosystems are extended to use LU decomposition with partial pivoting [62].

None of the aforementioned techniques use the novel structured deterministic phase mask in the hybrid transform domain using lower upper decomposition with partial pivoting. The motivation behind this paper is LU decomposition with partial pivoting [62] with an extension of hybrid transform. Here, a combination of DCT and FRT is used for asymmetric image encryption with different public and private keys. LUDPP technique is further used to achieve the private keys that are used in the decryption process. The public keys used are the two SDPM's that are designed using the combination of structured phase mask (SPM) and deterministic phase mask (DPM). To strengthen the cryptosystem hybrid transform is used along with the newly developed robust mask. The designed asymmetric cryptosystem is capable of enduring basic attacks on PTFT and is much secure and robust. Numerical simulations have been done for this system to prove the level of security.

This paper is composed as follows: Section 6.2 provides complete theoretical background for our proposed approach. Section 6.3 describes the proposed cryptosystem. Section 6.4 gives the simulation analysis to evaluate the security level through statistical analysis, attack analysis, number of pixels change rate (NPCR), and unified average change intensity (UACI) analysis. Section 6.5 represents the comparison with other related work. Finally, conclusions are made in Section 6.6.

## 6.2 Theoretical Background

### 6.2.1 Discrete Cosine Transform

The discrete cosine transform (DCT) is a mathematical transform that is used to compress a signal and transform it from spatial to the frequency domain. DCT representation of an image can be represented as a sum of sinusoids of varying magnitudes and frequencies. DCT is used in image compression applications because almost all the significant information about an image is concentrated in only a couple of coefficients of the DCT.

The two-dimensional DCT of an $M$-by-$N$ matrix $A$ can be defined as,

$$B_{pq} = \alpha_p \alpha_q \sum_{m=0}^{M-1} \sum_{n=0}^{N-1} A_{mn} \cos \frac{\pi(2m+1)p}{2M} \cos \frac{\pi(2n+1)q}{2N} \tag{6.1}$$

where $B_{pq}$ represents DCT coefficients of $A$, $p$ lies between 0 and $M - 1$ and $q$ lies between 0 and $N - 1$ and $\alpha_p$, $\alpha_q$ can be calculated as

$$\alpha_p = \begin{cases} 1/\sqrt{M}, p = 0 \\ \sqrt{2/M}, 1 \le p \le M - 1 \end{cases} \tag{6.2}$$

$$\alpha_q = \begin{cases} 1/\sqrt{N}, q = 0 \\ \sqrt{2/N}, 1 \le q \le N - 1 \end{cases} \tag{6.3}$$

The DCT is an invertible transform, and its inverse is given by

$$A_{mn} = \sum_{p=0}^{M-1}\sum_{q=0}^{N-1} \alpha_p\,\alpha_q\,B_{pq}\cos\frac{\pi(2m+1)p}{2M}\cos\frac{\pi(2n+1)q}{2N} \tag{6.4}$$

$$\alpha_p = \begin{cases} 1/\sqrt{M}, p = 0 \\ \sqrt{2/M}, 1 \le p \le M-1 \end{cases} \tag{6.5}$$

$$\alpha_q = \begin{cases} 1/\sqrt{N}, q = 0 \\ \sqrt{2/N}, 1 \le q \le N-1 \end{cases} \tag{6.6}$$

where $m$ lies between 0 and $M-1$ and $n$ lies between 0 and $N-1$.

### 6.2.2 Fractional Fourier Transform

The fractional Fourier transform (FRT) can be considered as a detailed version of the conventional Fourier transform (FT) with an additional component '$\alpha$' that is added to increase the strength of the traditional Fourier Transform. FRT is treated equivalent to the fundamental FT when the value of the order parameter $\alpha = 1$. FRT is widely used due to its capability of rotating a given signal at any desired angle, rather than just $\pi/2$ or its multiple while using FT. Similarly, FRT has some additional properties like the additive, inverse, and linearity that make it convenient to use for image encryption.

A kernel $K_\alpha$ [3] is used to map this transform which is also considered as an integral one with the help of the equations given below.

The fractional Fourier transform with order $\alpha$ for any signal $y(k)$ can be expressed as:

$$Y_\alpha(v) = \int_{-\infty}^{\infty} y(k)K_\alpha(k,v)\,dt \tag{6.7}$$

where the kernel of FRT is defined as:

$$K_\alpha(k,v) = \begin{cases} (k-v) & \text{if } \alpha \text{ is multiple of } 2\pi \\ (k+v) & \text{if } \alpha + \pi \text{ is multiple of } 2\pi \\ \sqrt{1-\frac{j\cot(\alpha)}{2\pi}}\,e^{j\left(\frac{v^2+k^2}{2}\right)\cot(\alpha)-jvk\,\mathrm{cosec}(\alpha)} & \text{if } \alpha \text{ is not a multiple of } \pi \end{cases} \tag{6.8}$$

and

$$Y_\alpha(v) = \begin{cases} \left(\sqrt{1-j\cot\frac{\alpha}{2\pi}}\,e^{j\left(v\cot\frac{\alpha}{2}\right)}\displaystyle\int_{-\infty}^{\infty} y(k)e^{j\left(k^2\cot\frac{\alpha}{2}\right)-juk\,\mathrm{cosec}(\alpha)}\,dt\right) & \text{if } \alpha \text{ is not a multiple of } \pi \\ y(k) & \text{if } \alpha \text{ is a multiple of } 2\pi \\ y(-k) & \text{if } \alpha + \pi \text{ is a multiple of } 2\pi \end{cases} \tag{6.9}$$

### 6.2.3 Structured Deterministic Phase Mask

To realize a structured deterministic phase mask (SDPM), we need several other components. When we combine these components, the result is a complex mask that is difficult to regenerate without accurate information of all the components.

Firstly, we need to generate the optical vortex phase mask (OVPM) which we get by multiplying two components i.e. Radial Hilbert Mask (RHM) and Fresnel Zone Plate as shown in Figure 6.1a, b. The OVPM is a kind of structured phase mask that has many advantages over the traditional RPM. Optical vortices even have the benefit of removing the axis alignment problem of optical setup, have attributes of different keys in a single mask that makes extra security parameters and are difficult to replicate [40]. The utilization of an OVPM rises the key space of the proposed cryptosystem.

The equation describing RHM [38] and Fresnel lens based on quadratic phase change are:

$$H = \exp(ip\theta) \text{ and } L(r) = \exp\left[-\frac{i\pi r^2}{\lambda f}\right] \tag{6.10}$$

Now, the OVPM [40] can be described as:

$$F_{\lambda, f}(r) = \exp\left[i\left(p\theta - \frac{\pi r^2}{\lambda f}\right)\right] \tag{6.11}$$

where $p$ is the topological charge, $f$ describes the focal length, $r$ describes the radius of the lens, $\lambda$ describes the wavelength of incident light. An OVPM is incorporated in the designed scheme to meet the high security and flexibility requirements and is shown in Figure 6.1c.

Secondly, we need to generate DPM. A DPM [45,46] generation requires defining of encryption order '$m$' which forms the basis of determining the number of subkeys (NSKs) NSK = $2^m \times 2^m$ in the mask. The value of encryption order can be $m = 2, 3, 4$. In the proposed scheme, for simplicity, we have used $m = 2$ for DPM generation. Then, we split the input image into NSK equal sub-blocks, each with size '$d$' which can be calculated as

$$d = \frac{dim}{2^m} \tag{6.12}$$

where *dim* represents the input image size.

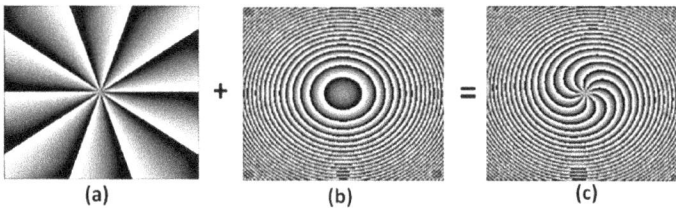

(a)     +     (b)     =     (c)

**FIGURE 6.1**
This figure shows the generation of Optical vortex phase mask by combining Radial Hilbert Mask and Fresnel Zone Plate.

DPM can now be generated using the continuous combination of sub masks ($M_{i,j}$) as represented below

$$\text{DPM} = \sum_{i=1}^{2^m}\sum_{j=1}^{2^m} M_{i,j}(d \times d) \tag{6.13}$$

where $M_{i,j}$ represents the number of sub blocks required for the generation of DPM. $M_{i,j}$ can be defined as:

$$M_{i,j}(p,q) = \exp\left[i2\pi\left(u_k.p + v_k \cdot q\right)\right] \tag{6.14}$$

where $k$ value lies in the interval $[1, 2^m]$, $u_k$ and $v_k$ are randomly generated within the range $[1, d]$.

Now, we finally generate the structured deterministic phase mask by taking the product of DPM and OVPM.

$$\text{SDPM} = \text{DPM} \times \text{OVPM} \tag{6.15}$$

Figure 6.2a, b represents the DPM and OVPM, which when combined form the SDPM that acts as a secret key in the proposed algorithm. SDPM1 and SDPM2 are shown in Figures 6.2c and 6.3c, respectively.

Ultimately, the eavesdropper task is made quite difficult by using structured deterministic phase mask i.e. the motivation behind the combination of different phase masks. The parameters to achieve the correct structured deterministic mask are quite large including, $r$, $\lambda$, $m$, $d$, $u_k$, $v_k$.

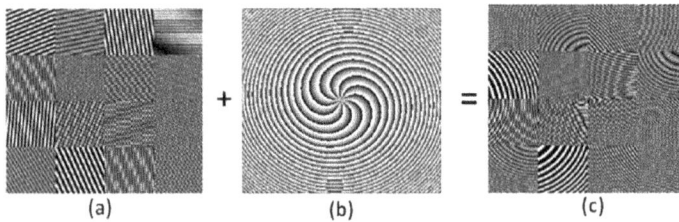

**FIGURE 6.2**
This figure shows the generation of Structured Deterministic Phase Mask 1 by combining Deterministic Phase Mask 1 and Optical Vortex Phase Mask.

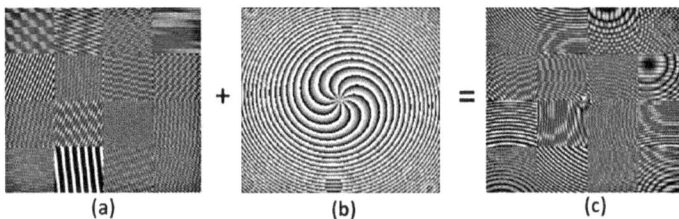

**FIGURE 6.3**
This figure shows the generation of Structured Deterministic Phase Mask 2 by combining Deterministic Phase Mask 2 and Optical Vortex Phase Mask.

## 6.3 Proposed Cryptosystem

The schematic diagram for the encryption and decryption process of the proposed system is given in Figure 6.4. $I(x,y)$ is the input image that is going to be encrypted. $(x,y)$ represents the indices of the input plane and $(u,v)$ represents the indices of the transform plane. SDPM1 and SDPM2 are the two structured deterministic masks that are the main keys for the encryption process. The encryption is carried out as explained below:

1) Firstly, the input image $I(x,y)$ is multiplied with the first structured deterministic phase mask SDPM1. A Discrete Cosine Transformation is performed on the $I(x,y) \times$ SDPM1 and then the transform spectrum is divided using LUDPP. This operation

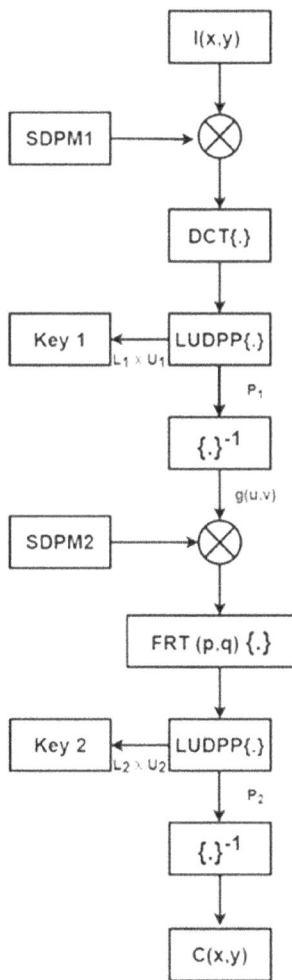

**FIGURE 6.4**

This figure depicts the whole encryption flowchart, i.e., how we get a cipher image from the original image after applying the encryption algorithm and encryption keys.

outputs the first private key Key 1 and generates the intermediate output $g(u,v)$ represented using the equations below:

$$[L_1,U_1,P_1] = \text{LUDPP}\{\text{DCT}[I(x,y)\times\text{SDPM1}]\} \tag{6.16}$$

$$\text{Key 1} = L_1 \times U_1 \tag{6.17}$$

$$g(u,v) = P_1^{-1}(u,v) \tag{6.18}$$

where LUDPP{.} express the lower upper decomposition with partial pivoting, $L_1$, $U_1$ and $P_1$ are the three products of first LUDPP operation, DCT{.} expresses the Discrete Cosine Transform, × represents the matrix multiplication operation, {.}$^{-1}$ denotes the inverse matrix operation.

2) The intermediate output $g(u,v)$ is then multiplied by second structured deterministic mask SDPM2. The fractional Fourier transform FRT$(p,q)$ is applied on this resultant followed by second LUDPP which gives the output as second private key Key 2 and the final cipher image denoted by $C(x,y)$. The equations are given below:

$$[L_2,U_2,P_2] = \text{LUDPP}\{\text{FRT}(p,q)[g(u,v)\times\text{SDPM2}]\} \tag{6.19}$$

$$\text{Key 2} = L_2 \times U_2 \tag{6.20}$$

$$C(x,y) = P_2^{-1}(x,y) \tag{6.21}$$

where $L_2$, $U_2$ and $P_2$ are the three products of second LUDPP operation, FRT$(p,q)$much{.} expresses the fractional Fourier Transform. The use of inverse matrix of the two-permutation matrix in the proposed scheme results the cipher image.

This is the complete process of encryption which generates two private keys namely, Key 1 and Key 2 which will be used in the decryption process.

The decryption algorithm is not similar to encryption one as the system we proposed is the asymmetric one. The process of decryption is carried out as follows when we have the two private keys generated during the encryption step:

1) The cipher image $C(x,y)$ that is produced as final output of encryption process is taken as an input during the decryption process. Firstly, it is multiplied with the private key Key 2 and the resultant function is then subjected to inverse fractional Fourier transform with order parameters $(-p,-q)$. The result of this transformation can be represented as $g(u,v)$ calling it an intermediate matrix given by:

$$g(u,v) = \text{abs}\{\text{FRT}(-p,-q)[C(x,y)\times\text{Key 2}]\} \tag{6.22}$$

where *abs* defines the amplitude part extraction of a complex function.

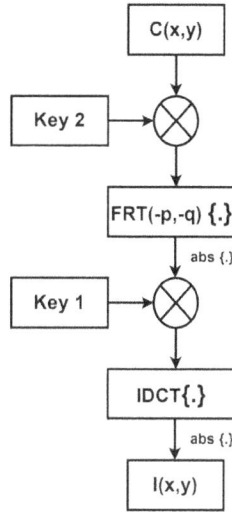

**FIGURE 6.5**
This figure depicts the decryption flowchart, i.e., how we get the original image back from the cipher image after application of correct decryption algorithm and decryption keys.

2) To obtain the final decrypted image, the intermediate matrix $g(u,v)$ is multiplexed with the private key Key 1 and then subject it to the inverse discrete cosine transform. The decrypted image can be given by:

$$I(x,y) = abs\{\text{IDCT}\left[g(u,v) \times \text{Key }1\right]\} \qquad (6.23)$$

In the encryption process, the permutation matrix $P$ inverse is taken and recorded as the intermediate result or the final cipher image, which has two points of interest. Firstly, $P^{-1}$ is a sparse matrix which means its most of the elements are zero. This property makes the cipher text resistant to iterative attack based on amplitude phase retrieval technique despite the fact that the attacker knows the original ciphertext. Thus, it will be a challenging task to recover the plaintext from the known ciphertext with no information of private keys. It makes the cryptosystem more secure and robust. Furthermore, $P^{-1}$ is a matrix having real values which make it easy to be directly recorded by charged coupled devices (CCD). Thus, no hologram technique is needed to acquire the ciphertext. Hence, the proposed scheme is simple yet secure. As seen, the encryption keys (public keys) are different from the two recorded private keys (Key1 and Key 2) and the process of encryption and decryption are also different as shown in Figures 6.4 and 6.5. Thus, it can be said that the above proposed cryptosystem is asymmetric in nature.

Figure 6.6 shows the optical execution of our designed decryption algorithm. To regulate both the amplitude and phase information two spatial light modulators (SLMs), SLM1 and SLM2, are utilized. PC is the personal computer used to carry out the matrix multiplication, such as $c \times$ Key 2 and $g \times$ Key 1. The final decrypted image is acquired using a CCD.

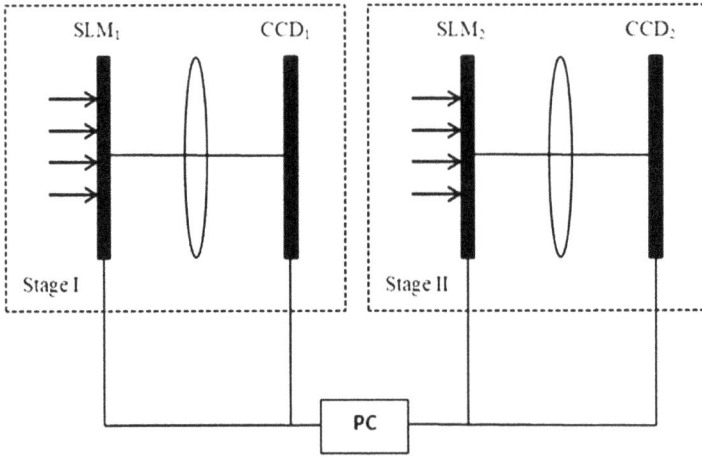

**FIGURE 6.6**
This figure shows the optical setup for decryption, i.e., how the decryption process of the proposed cryptosystem can be performed optically.

## 6.4 Simulation and Evaluation

The effectiveness and feasibility of the presented technique have been checked by carrying out various numerical simulations and analysis. Matlab has been used for the simulation purpose. Two grayscale images (Cameraman and Peppers, shown in Figure 6.7a and b) of size 256 × 256 are taken as the input images to be encrypted using the proposed technique. Figure 6.7c, d shows the first set of private keys generated for both the input images employing the proposed algorithm and Figure 6.7e, f shows the second set of private keys respectively for Cameraman and Peppers images. The final encrypted images are depicted in Figure 6.7g, h post the encryption process. When the decryption process is correctly carried out with the right set of decryption keys, the decrypted images are achieved (Figure 6.7I, j) which are similar to the original input images.

### 6.4.1 Statistical Analysis

To figure out the authenticity and strength of the proposed technique, mean square error (MSE), and peak signal to noise ratio (PSNR) has been calculated for the decrypted image $I'(x,y)$ and the original image $I(x,y)$ [38].

$$\text{MSE} = \sum_{x=1}^{N}\sum_{y=1}^{N} \frac{\left|I'(x,y) - I(x,y)\right|^2}{M \times N} \tag{6.24}$$

where $M \times N$ represents the size of the image.

$$\text{PSNR} = 10 \times \log_{10}\left(\frac{(N-1)^2}{\text{MSE}}\right) \tag{6.25}$$

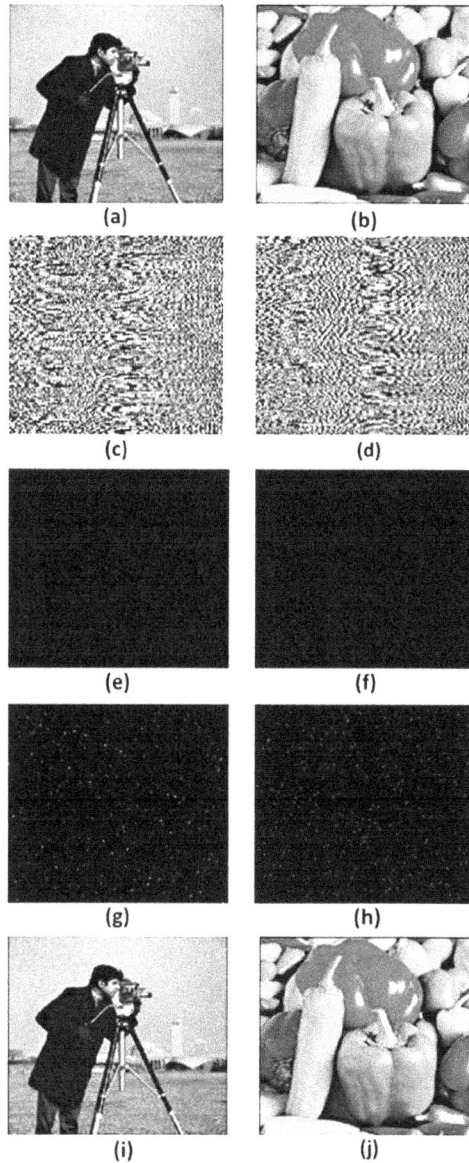

**FIGURE 6.7**
This figure shows the simulation result of the proposed cryptosystem including Original Cameraman and Peppers images, both Private keys, Encrypted images, and Decrypted images.

where $N$ represents the maximum number of pixels in the original and decrypted images. Here the value of $N$ is 256 as the size of images are $256 \times 256$.

The calculate value of MSE for Cameraman image is $9.0699 \times e^{-25}$ and for peppers image is $6.4786 \times e^{-25}$. PSNR value for Cameraman image is 288.558 dB and for peppers image is 290.016 dB Lower the value of MSE and higher the value of PSNR represents the competency of the proposed method.

## 6.4.2 Attack Analysis

Occlusion attack analysis has been done for the encrypted images and the masks. As it is known, the greater parts of the occlusion examination are done uniquely for encrypted images in the past. Here, in this work the occlusion analysis is done for both encrypted images and the new mask developed. The occlusion examination for encrypted images only proves the toughness of the cryptosystem while if it is done for the newly developed structured deterministic phase mask, it confirms the stability and robustness of the cryptosystem as well as the mask. Consequently, to check the strength of the proposed framework, power of SDPM is approved by performing occlusion examination on SDPM. Few portions of encrypted images are occluded in Figure 6.8, and examinations are done on decrypted ones. Twenty-five percent (25%) occluded encrypted images are shown in Figure 6.8a for Cameraman and Figure 6.8c for Peppers. Their respective decrypted images are shown in Figure 6.8b, d respectively. Similarly, 43.75% occluded encrypted images for Cameraman and Peppers are shown in Figure 6.8e, g. Corresponding decrypted images are shown in Figure 6.8f, h, respectively, for Cameraman and Peppers. The effect of occlusion on the encrypted image shows that the decrypted image is identifiable with respect to the original one even if a small part of the encrypted image is lost due to some kind of network failure or uncoordinated transmission of data. Hence, it shows the strength of the scheme that even if some part of the data is missing during transmission, the receiver can still get most of the information in the decrypted result. It proves the scheme is resistant to a certain degree of occlusion.

Similarly, when the effect of occlusion is checked on the mask SDPM, some part of the mask is occluded and examinations are performed on decrypted images. Figure 6.9a–c shows the 23.4%, 43.75%, 75% occluded part on the mask SDPM for Cameraman image. The equivalent decrypted images are depicted in Figure 6.9d–f. Likewise, Figure 6.9g–i denotes 23.4%, 43.75%, 75% occluded part on the mask SDPM for Peppers image and corresponding decrypted images are shown in Figure 6.9j–l.

Execution estimates like MSE and PSNR are additionally determined for decrypted images after occlusion. Table 6.1 shows the value of MSE and PSNR for Cameraman and

**FIGURE 6.8**
This figure shows the 25% and 43.75% occluded encrypted images for Cameraman and Peppers and their corresponding decrypted images.

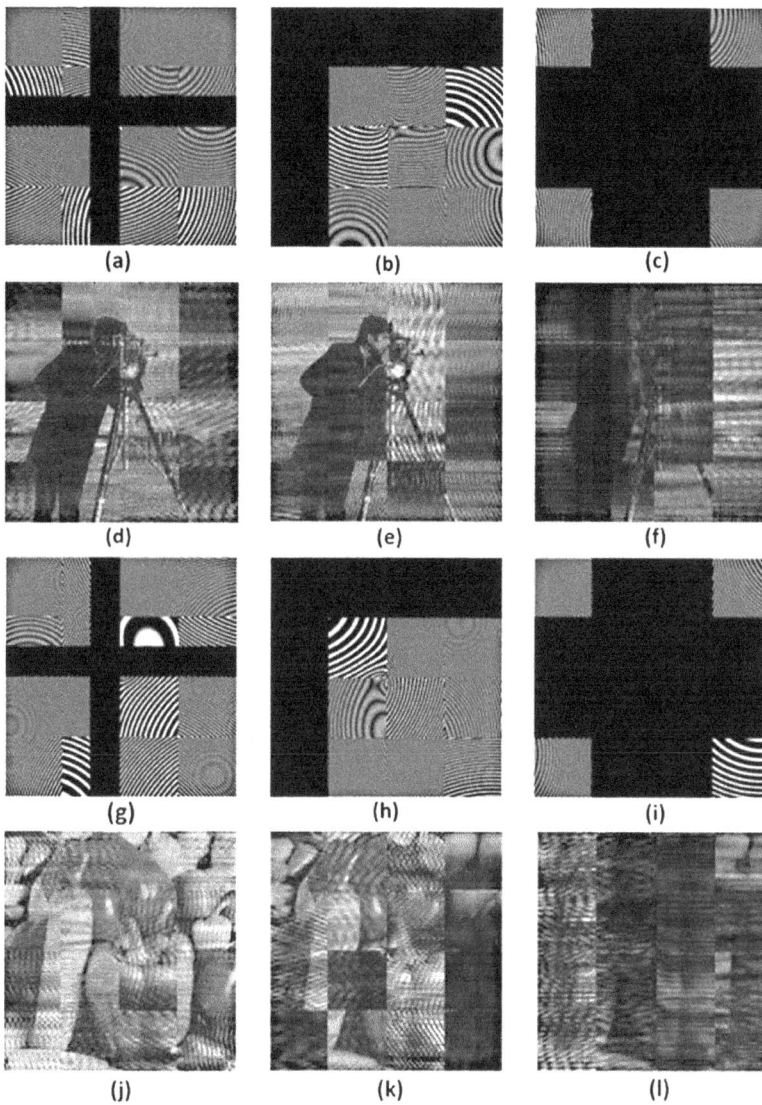

**FIGURE 6.9**
This figure shows the effect of occlusion on the structured deterministic phase mask. It shows 23.4%, 43.75%, 75% occluded SDPM for Cameraman and Peppers with their corresponding decrypted images.

**TABLE 6.1**

A Table Comparing the MSE and PSNR Values for Decrypted Cameraman and Peppers Image When Different Percentages of Mask is Occluded

| | 23.4% Occlusion | | 43.75% Occlusion | | 75% Occlusion | |
|---|---|---|---|---|---|---|
| | MSE | PSNR | MSE | PSNR | MSE | PSNR |
| Cameraman | $3.5139 \times 10^5$ | $-7.3271$ | $3.0359 \times 10^5$ | $-6.6921$ | $1.4824 \times 10^6$ | $-13.5789$ |
| Peppers | $4.3070 \times 10^4$ | $-1.7891$ | $6.0963 \times 10^5$ | $-9.7199$ | $1.1543 \times 10^6$ | $-12.6916$ |

Peppers images when there is 23.4% occlusion, 43.75% occlusion, and 75% occlusion in the structured deterministic mask.

### 6.4.3 NPCR and UACI Analysis

In order to check the resistance of the proposed scheme to differential attacks, these analyses have been done. NPCR represents the number of pixels change rate and UACI stands for unified average change intensity. These calculations are used to observe small changes but if the result is a big change after encryption, then the cryptosystem is attack resistant. The relationship between the original and encrypted image is built up utilizing NPCR parameter and UACI parameter is used to determine an average density between two images $I(i, j)$ and $C(i, j)$ [63]. The formulas for the computation of these parameters are given in the equation below. For calculating NPCR value we need to first determine the value of $D(i,j)$. If the pixel estimations of two images are unique, at that point $D(i,j)$ is 1, else it is 0. $M \times N$ represents the pixel of the image.

$$\mathrm{NPCR}\left(I,C\right)=\frac{\Sigma_{i,j}D\left(i,j\right)}{M\times N}\times100\%I\left(i,j\right) \qquad (6.26)$$

$$\left(i,j\right)=\begin{cases}1,\text{if }I\left(i,j\right)\neq C\left(i,j\right)\\0,\text{if }I\left(i,j\right)=C\left(i,j\right)\end{cases} \qquad (6.27)$$

$$\mathrm{UACI}\left(I,C\right)=\Sigma_{i,j}\frac{\left|I\left(i,j\right)-C\left(i,j\right)\right|}{M\times N\times255}\times100\% \qquad (6.28)$$

The ideal value of these two parameters NPCR and UACI is 99.61 and 33.4%. Table 6.2 below gives the comparative analysis of NPCR and UACI for [53,63] and the proposed scheme. NPCR and UACI estimations of the proposed framework are 0.99221 and 35.27% and these values are near perfect. Subsequently, the proposed framework gives great outcomes and it is safe against differential attacks.

### 6.4.4 Key Sensitivity Analysis

Here, we explore the scheme sensitivity against different encryption parameters. Different wrong combinations have been implemented to check the robustness of the proposed method and to validate the strength of the scheme. Figure 6.10a, b depicts the decrypted output when we swap the fractional order while performing decryption for the Cameraman

**TABLE 6.2**

A Table Comparing the NPCR and UACI Values of Different Schemes with the Proposed One

| Algorithm | NPCR | UACI |
|---|---|---|
| Ref. [53] | 0.98215 | 0.3231 |
| Ref. [63] | 0.95500 | 0.3214 |
| Proposed Scheme | 0.99221 | 0.3527 |

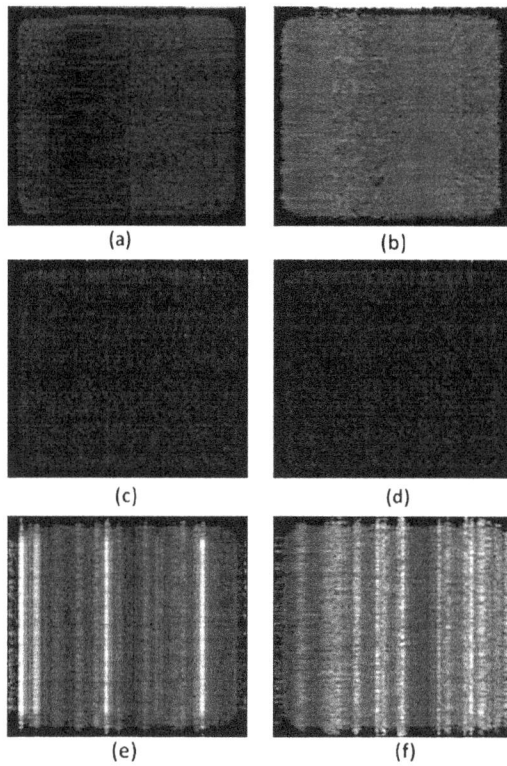

**FIGURE 6.10**
This figure shows the key sensitivity analysis means what is the effect of wrong keys on the decryption algorithm. Like when the fractional orders are exchanged, when the private keys are exchanged, when the order of applying the transforms is changed while decryption for Cameraman and Peppers respectively.

and Peppers image respectively. When we swap the private decryption keys during decryption, the decrypted output has been seen in Figure 6.10c, d for. Similarly, while performing decryption when we swap the order of applying the transform is changed, the output is shown in Figure 6.10e, f for both the Cameraman and Peppers images respectively. All the decrypted output shows that with any deviation in the encryption parameters, we are not able to get an idea about what is the original image. So, it can be analyzed from the results that the scheme is sensitive to any change in the encryption keys.

## 6.5 Performance Comparison

The comparison of performance of the proposed algorithm with some of the widely known image encryption algorithms has been done based on various parameters. The Table 6.3 shows the comparison and strengthens the fact that the performance of the proposed scheme is much better than the previous ones. By practicing the hybrid transform, the advantage of increase in the security parameters is achieved.

**TABLE 6.3**

A Table Comparing the Performance Based on Various Parameters of Proposed Algorithm with Other Encryption Techniques

| Parameters | DRPE with Fractional Fourier Transform | LUD with Fractional Fourier Transform and Arnold Transform | LUD with Partial Pivoting and Two Random Phase Mask | Proposed Technique with PPLUD, SDPM Using DCT & FRT |
|---|---|---|---|---|
| Transform used | Fractional Fourier | Fractional Fourier and Arnold | Fourier | Discrete Cosine and Fractional Fourier |
| Applied Technique | Optical and Digital | Digital | Optical and Digital | Optical and Digital |
| Number of Diffusers | 2 RPM | Random Amplitude Mask and RPM | 2 RPM | 2 SDPM |
| Number of Security Parameters | Fractional Order | Fractional Order and Arnold Parameter | Nil | 2 fractional orders, $p, f, r, \lambda, m, d, u_k, v_k$ |
| Execution Time (seconds) | 0.2018 | 0.3139 | 0.2836 | 0.6438 |
| Strength | Low Complexity and Fast | Endurance to chosen plaintext, known plaintext, cipher text only attacks | Several attack free cryptosystem, PT process is replaced by partial pivoting | Enhanced security potential by using SDPM, additional parameters of hybrid transform, and robustness against attacks |
| Weakness | Easy to Break, Prone to Attack | Simple and common masks used, which are easy to do | Small and simple key space | A bit more execution time |

## 6.6 Conclusion

A novel structured deterministic phase mask along with lower upper decomposition with partial pivoting in hybrid transform domain has been utilized in the designing of the asymmetric cryptosystem. The main goal of the proposed scheme is to provide greater security to the LUDPP based encryption technique and to make it more confidential and robust. The proposed strategy has several features; firstly, it is influential and robust due to the introduction of a secure novel structured deterministic phase mask which is difficult to realize without accurate information of all the parameters used in its construction. Secondly, it is fast and the encryption process can be realized digitally while decryption can be performed optically. Thirdly, it is an asymmetric cryptosystem where the two SDPMs are considered as public keys and the private (decryption) keys are generated by LUDPP during the encryption process which generates unique plain text that helps in resisting known-plaintext attacks. Lastly, the cipher text matrix generated by partial pivoting is a sparse matrix, which resists iterative attacks as it does not provide enough constraints to make it stable. The scheme has been verified on different factors, for example, MSE, PSNR, NPCR, UACI to check its adequacy and strength. It has been analyzed that even under distortions caused by occlusion, one is able to recuperate the original image. The key space of the presented scheme is effectively large and the simulation results confirm the robustness and efficacy of the proposed cryptosystem.

## Declarations

- No funding was received for conducting this study.
- The authors have no conflicts of interest to declare that are relevant to the content of this article.
- The authors have no relevant financial or non-financial interests to disclose.

## References

1. Refregier, P. and Javidi, B. Optical image encryption based on input plane and Fourier plane random encoding. *Optics Letters* 20(7), 767–769 (1995).
2. Unnikrishnan, G. and Singh, K. Double random fractional Fourier domain encoding for optical security. *Optical Engineering* 39(11), 2853–2860 (2000).
3. Unnikrishnan, G., Joseph, J. and Singh, K. Optical encryption by double-random phase encoding in the fractional Fourier domain. *Optics Letters* 25(12), 887–889 (2000).
4. Situ, G. and Zhang, J. Double random-phase encoding in the Fresnel domain. *Optics Letters* 29(14), 1584–1586 (2004).
5. Chen, L. and Zhao, D. Optical image encryption with Hartley transforms. *Optics Letters* 31(23), 3438–3440 (2006).
6. Rodrigo, J.A., Alieva, T. and Calvo, M.L. Gyrator transform: Properties and applications. *Optics Express* 15(5), 2190–2203 (2007).
7. Matoba, O., Nomura, T., Perez-Cabre, E., Millan, M.S. and Javidi, B. Optical techniques for information security. *Proceedings of the IEEE* 97(6), 1128–1148 (2009).
8. Peng, X., Zhang, P., Wei, H. and Yu, B. Known-plaintext attack on optical encryption based on double random phase keys. *Optics Letters* 31(8), 1044–1046 (2006).
9. Rajput, S.K. and Nishchal, N.K. Known-plaintext attack on encryption domain independent optical asymmetric cryptosystem. *Optics Communications* 309, 231–235 (2013).
10. Gopinathan, U., Monaghan, D.S., Naughton, T.J. and Sheridan, J.T. A known-plaintext heuristic attack on the Fourier plane encryption algorithm. *Optics Express* 14(8), 3181–3186 (2006).
11. Carnicer, A., Montes-Usategui, M., Arcos, S. and Juvells, I. Vulnerability to chosen-cyphertext attacks of optical encryption schemes based on double random phase keys. *Optics Letters* 30(13), 1644–1646 (2005).
12. Zhang, Y., Xiao, D., Wen, W. and Liu, H. Vulnerability to chosen-plaintext attack of a general optical encryption model with the architecture of scrambling-then-double random phase encoding. *Optics Letters* 38(21), 4506–4509 (2013).
13. Peng, X., Wei, H. and Zhang, P. Chosen-plaintext attack on lensless double-random phase encoding in the Fresnel domain. *Optics Letters* 31(22), 3261–3263 (2006).
14. Kumar, P., Kumar, A., Joseph, J. and Singh, K. Vulnerability of the security enhanced double random phase-amplitude encryption scheme to point spread function attack. *Optics and Lasers in Engineering* 50(9), 1196–1201 (2012).
15. Kumar, P., Joseph, J. and Singh, K. Double random phase encoding based optical encryption systems using some linear canonical transforms: Weaknesses and countermeasures. *In Linear Canonical Transforms* 198, 367–396 (2016).
16. Jiao, S., Li, G., Zhou, C., Zou, W. and Li, X. Special ciphertext-only attack to double random phase encryption by plaintext shifting with speckle correlation. *Journal of the Optical Society of America A* 35(1), A1–A6 (2018).
17. Liu, X., Wu, J., He, W., Liao, M., Zhang, C. and Peng, X. Vulnerability to ciphertext-only attack of optical encryption scheme based on double random phase encoding. *Optics Express* 23(15), 18955–18968 (2015).

18. Li, G., Yang, W., Li, D. and Situ, G. Cyphertext-only attack on the double random-phase encryption: experimental demonstration. *Optics Express* 25(8), 8690–8697 (2017).
19. Rajput, S.K. and Nishchal, N.K. Fresnel domain nonlinear optical image encryption scheme based on Gerchberg–Saxton phase-retrieval algorithm. *Applied Optics* 53(3), 418–425 (2014).
20. Hwang, H.E., Chang, H.T. and Lie, W.N. Multiple-image encryption and multiplexing using a modified Gerchberg-Saxton algorithm and phase modulation in Fresnel-transform domain. *Optics Letters* 34(24), 3917–3919 (2009).
21. Hwang, H.E., Chang, H.T. and Lie, W.N. Fast double-phase retrieval in Fresnel domain using modified Gerchberg-Saxton algorithm for lensless optical security systems. *Optics Express* 17(16), 13700–13710 (2009).
22. Wang, Q., Guo, Q. and Lei, L. Multiple-image encryption system using cascaded phase mask encoding and a modified Gerchberg–Saxton algorithm in gyrator domain. *Optics Communications* 320, 12–21 (2014).
23. Liu, X., Cao, Y., Lu, P., Lu, X. and Li, Y. Optical image encryption technique based on compressed sensing and Arnold transformation. *Optik* 124(24), 6590–6593 (2013).
24. Liu, X., Mei, W. and Du, H. Optical image encryption based on compressive sensing and chaos in the fractional Fourier domain. *Journal of Modern Optics* 61(19), 1570–1577 (2014).
25. Zhou, N., Zhang, A., Wu, J., Pei, D. and Yang, Y. Novel hybrid image compression–encryption algorithm based on compressive sensing. *Optik* 125(18), 5075–5080 (2014).
26. Chen, J., Zhang, Y., Qi, L., Fu, C. and Xu, L. Exploiting chaos-based compressed sensing and cryptographic algorithm for image encryption and compression. *Optics & Laser Technology* 99, 238–248 (2018).
27. Zang, J., Xie, Z. and Zhang, Y. Optical image encryption with spatially incoherent illumination. *Optics Letters* 38(8), 1289–1291 (2013).
28. Qin, W. and Peng, X. Asymmetric cryptosystem based on phase-truncated Fourier transforms. *Optics Letters* 35(2), 118–120 (2010).
29. Qin, W., Peng, X., Gao, B. and Meng, X. Universal and special keys based on phase-truncated Fourier transform. *Optical Engineering* 50(8), 080501 (2011).
30. Chen, W. and Chen, X. Optical color image encryption based on an asymmetric cryptosystem in the Fresnel domain. *Optics Communications* 284(16–17), 3913–3917 (2011).
31. Mehra, I. and Nishchal, N.K. Optical asymmetric image encryption using gyrator wavelet transform. *Optics Communications* 354, 344–352 (2015).
32. Wang, Q., Guo, Q. and Zhou, J. Color image hiding based on phase-truncation and phase retrieval technique in the fractional Fourier domain. *Optik* 124(12), 1224–1229 (2013).
33. Wang, X. and Zhao, D. A special attack on the asymmetric cryptosystem based on phase-truncated Fourier transforms. *Optics Communications* 285(6), 1078–1081 (2012).
34. Wang, X. and Zhao, D. Security enhancement of a phase-truncation based image encryption algorithm. *Applied Optics* 50 (36), 6645–6651 (2011).
35. Rajput, S.K. and Nishchal, N.K. Image encryption using polarized light encoding and amplitude and phase truncation in the Fresnel domain. *Applied Optics* 52 (18), 4343–4352 (2013).
36. Mehra, I. and Nishchal, N.K. Image fusion using wavelet transform and its application to asymmetric cryptosystem and hiding. *Optics Express* 22(5), 5474–5482 (2014).
37. Wang, X. and Zhao, D. Double images encryption method with resistance against the specific attack based on an asymmetric algorithm. *Optics Express* 20(11), 11994–12003 (2012).
38. Maan, P. and Singh, H. Non-linear cryptosystem for image encryption using radial Hilbert mask in fractional Fourier transform domain. 3D *Research* 9(4), 53 (2018).
39. Maan, P., Singh, H. and Kumari, A.C. Image encryption based on Walsh Hadamard and fractional Fourier transform using Radial Hilbert Mask. *International conference on computing and communication technologies for smart nation (IC3TSN)*, Gurgaon, India. pp. 179–183 (2017).
40. Singh, H. Nonlinear optical double image encryption using random-optical vortex in fractional Hartley transform domain. *Optica Applicata* 47(4), 557–578 (2017).
41. Maan, P., Singh, H. and Kumari, A.C. Optical asymmetric cryptosystem based on Kronecker product, hybrid phase mask and optical vortex phase masks in the phase truncated hybrid transform domain 3D *Research* 10(1), 8 (2019).

42. Zamrani, W., Ahouzi, E., Lizana, A., Campos, J. and Yzuel, M.J. Optical image encryption technique based on deterministic phase masks. *Optical Engineering* 55(10), 103108 (2016).
43. Rajput, S.K. and Nishchal, N.K. Asymmetric color cryptosystem using polarization selective diffractive optical element and structured phase mask. *Applied Optics* 51(22), 5377–5386 (2012).
44. Liansheng, S., Bei, Z., Xiaojuan, N. and Ailing, T. Optical multiple-image encryption based on the chaotic structured phase masks under the illumination of a vortex beam in the gyrator domain. *Optics Express* 24(1), 499–515 (2016).
45. Liu, L. and Miao, S. A new image encryption algorithm based on logistic chaotic map with varying parameter. *SpringerPlus* 5(1), 289 (2016).
46. Rakheja, P., Vig, R. and Singh, P. Double image encryption using 3D Lorenz chaotic system, 2D non-separable linear canonical transform and QR decomposition. *Optical and Quantum Electronics* 52(2), 103 (2020).
47. Chen, J., Zhang, Y., Li, J. and Zhang, L.B. Security enhancement of double random phase encoding using rear-mounted phase masking. *Optics and Lasers in Engineering* 101, 51–59 (2018).
48. Javidi, B., Carnicer, A., Yamaguchi, M., Nomura, T., Pérez-Cabré, E., Millán, M.S., Nishchal, N.K., Torroba, R., Barrera, J.F., He, W. and Peng, X. Roadmap on optical security. *Journal of Optics* 18(8), 083001 (2016).
49. Wang, X. and Zhao, D. Amplitude-phase retrieval attack free cryptosystem based on direct attack to phase-truncated Fourier-transform-based encryption using a random amplitude mask. *Optics Letters* 38(18), 3684–3686 (2013).
50. Liansheng, S. and Zhanmin, W. Amplitude-phase retrieval attack free image encryption based on two random masks and interference. *Optics and Lasers in Engineering* 86, 1–10(2016).
51. Xiong, Y., He, A. and Quan, C. Hybrid attack on an optical cryptosystem based on phase-truncated Fourier transforms and a random amplitude mask. *Applied Optics* 57(21), 6010–6016 (2018).
52. Xiong, Y., He, A. and Quan, C. Specific attack and security enhancement to optical image cryptosystem based on two random masks and interference. *Optics and Lasers in Engineering* 107, 142–148 (2018).
53. Khurana, M. and Singh, H. An asymmetric image encryption based on phase truncated hybrid transform. *3D Research* 8(3), 1–17 (2017).
54. Tao, S., Tang, C., Shen, Y. and Lei, Z. Optical image encryption based on biometric keys and singular value decomposition. *Applied Optics* 59(8), 2422–2430 (2020).
55. Su, Y., Xu, W. and Zhao, J. Optical image encryption based on chaotic fingerprint phase mask and pattern-illuminated Fourier ptychography. *Optics and Lasers in Engineering* 128, 106042 (2020).
56. Zhu, Z., Wu, C., Wang, J., Hu, K. and Chen, X.D. A novel 3D vector decomposition for color-image encryption. *IEEE Photonics Journal* ID: 216287602 (2020).
57. Shan, M., Liu, L., Liu, B. and Zhong, Z. Asymmetric multiple-image encryption based on equal modulus decomposition and frequency modulation. *Laser Physics* 30(3), 035202 (2020).
58. Khurana, M. and Singh, H. Two level phase retrieval in fractional Hartley domain for secure image encryption and authentication using digital signatures. *Multimedia Tools and Applications* 19–20, 13967–13986 (2020).
59. Abdelfattah, M., Hegazy, S.F., Areed, N.F. and Obayya, S.S. Compact optical asymmetric cryptosystem based on unequal modulus decomposition of multiple color images. *Optics and Lasers in Engineering* 129, 106063 (2020).
60. Xiong, Y., Du, J. and Quan, C. Optical encryption and authentication scheme based on phase-shifting interferometry in a joint transform correlator. *Optics & Laser Technology* 126, 106108 (2020).
61. Abuturab, M.R. Single-channel color information security system using LU decomposition in gyrator transform domains. *Optics Communications* 323, 100–109 (2014).
62. Xiong, Y. and Quan, C. Hybrid attack free optical cryptosystem based on two random masks and lower upper decomposition with partial pivoting. *Optics & Laser Technology* 109, 456–464 (2019).
63. Girija, R. and Singh, H. Enhancing security of double random phase encoding based on random S-Box. *3D Research* 9(2), 15(2018).

# 7

# *A Comparative Analysis of Blockchain Integrated IoT Applications*

Seema Verma

*Delhi Technical Campus, Greater Noida, India*

## CONTENTS

## 7.1  Introduction

In the current scenario, the things used by human beings (things in our surroundings) are expected to be worked on the internet. All the physical things which human beings can think of are going to be interconnected through the internet. The physical devices may

DOI: 10.1201/9781003206088-7

include lights, ACs, washing machines, refrigerators, and maybe even watches, house keys, toothbrushes, and many more in our surroundings; the name reforms from the Internet of Computers to Internet of Things (IoT). This exponential connectivity growth may be in billions and trillions of things to be interconnected; the concept of IoT is the foundation for this interconnection (Atzori et al., 2010; Bi et al., 2014). Unlike the network of computers, the network of physical devices is based on different configurations, specifications; they may be based on Cloud, Big Data, electrical science, mechanical science, and many other diverse technologies. The different connectivity for the devices is RFIDs, Wi-Fi, Zigbee, Cellular connectivity, Lowpan, Lora, and many others.

Currently nano devices are also thought to be the part of the Internet of (nano) Things, where the nano devices are expected to communicate with each other; e.g., nano capsule is to be swallowed and made to function inside the body; after functioning it is be excreted and communicated with other nano devices. In the shared economy, Airbnb, Uber etc. are the best examples.

The different applications are business and manufacturing, healthcare, retail sector, vehicular systems, security (biometric), smart locks, smart dust (chemical in the soil or diagnose problems in the human body) and many others we can think about, like air pollution, data exchange in oil platform considering the weather conditions, water leakages, snow level monitoring, earthquake pre-detection. Various challenges are faced by the IoT environment like exponential growth of devices, security, privacy, interoperability standards, regulatory issues, economy, and development issues; among these, security is the major concern when a trust-less distributed environment is in use.

Chapter focuses on the IoT security challenge with Blockchain technology with the flow as: IoT architecture and challenges are addressed in the second section followed by the introduction of Blockchain technology with its technical background in the next section. In the fourth section, the integration of IoT applications is discussed with Blockchain technology. Analysis of various applications is shown in the fifth section. The chapter concludes in the sixth section with its final remarks.

## 7.2 Introduction to Internet of Things

Like computers and limited computing devices are connected through the internet, more and more devices (physical devices) are going to be interconnected through the internet. This may ultimately lead to exponential growth to the internet. Concept of IoT is the foundation for this interconnection with many challenges (Alqassem & Svetinovic, 2014; Atzori et al., 2010; Bi et al., 2014; Evans, 2011; Miorandi et al., 2012; O'Halloran et al., 2015; Pereira et al., 2017; Sadeghi et al., 2015; Schneider, 2017; Wollschlaeger et al., 2017). One of the big challenges is addressing a very large number of devices (say trillions). These things can be connected to the existing network, thus by increasing the number of nodes in the current network. Another way can be to separate these things from the existing network, thus by making a new network without disturbing the existing network. Both the approaches are having their own challenges. Another problem is mobility, e.g., smart watch; it moves whenever the person moves. Nodes may change the underlying network. Various connectivity standards are available e.g., Wi-Fi, Ethernet, Bluetooth. Another big challenge is device security. One important aspect is identity, i.e., the device is the correct one. After identity, everything must be tamperproof, which means integrity must be maintained.

### 7.2.1 IoT Architecture

IoT architecture, in Figure 7.1, is presented briefly for IoT enterprises requiring the following components:

> Level 1 mainly includes the devices with sensors and actuators to collect, process, and generate the signals. For example, a smartphone senses the gravitational pull of the earth and the data is processed enough to orient the display in landscape view. Thus, the processing is done at this stage as per the capability of the device itself. Actuators can change the physical condition of the device that generated the data, e.g., the boiler's on/off switch, AC temperature etc.

The analogous data received at level 1 is received and aggregated from various resources. Then this aggregated data is converted to digital form by data acquisition system of level 2. This aggregated system is in close proximity to level 1 and further sent to the next level for the processing.

Before going to the centralized system, the accumulated data is pre-processed at level 3. There are many sensors which will send the data to an IoT device, say in TeraBytes (TB), this amount of data if sent directly, the data center will create chaos. That is why it's a good approach to pre-process this bulky data to generate meaningful results before going to the data center and sending this meaningful data only to the data center.

Finally, data is processed in a data center for in-depth processing. Level 4 is used only when immediate action is not required. As this level is time-consuming, critical operations are included, e.g., surgery, car crash etc. where immediate action is required this level is not included.

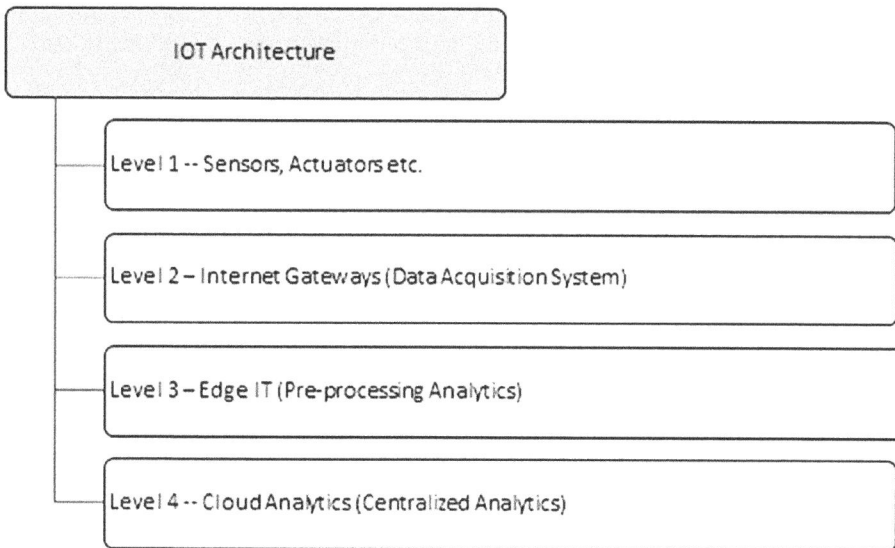

**FIGURE 7.1**
A rectangular shape at the top depicting IoT architecture with its four subparts of rectangular shape depicting four levels of IoT architecture.

### 7.2.2 Challenges in IoT

There are many challenges in IoT systems (Miorandi et al., 2012; Pereira et al., 2017; Sadeghi et al., 2015). A few are mentioned here:

Security: The security of IoT devices is a must. The insecure devices can be acting as loopholes for attacking the entire network.

Privacy: As the devices are connected to the internet, personal data can be used and analyzed by the company or government. The usage of personal data can later be misused by anyone.

Interoperability standards: Information must be exchanged between all the devices which are interconnected. But the difference in their complexity of underlying communication protocols makes it difficult.

Regulatory issues: The interconnected devices face many security issues, but there are no firm laws that cover all the IoT levels.

Economy and development issues: Though the low cost of hardware (microprocessors, sensors, actuators) can make IoT feasible in low economic areas. There are many other issues like the establishment of high-speed internet and other IoT technology architecture which hinder the development of IoT.

## 7.3 Introduction to Blockchain

Blockchain (Nakamoto, 2008) is used for coordinating and collaborating in an environment where participants do not trust each other and hence the system cannot be set up in a centralized manner. Such type of distributed environment is illustrated in the figure (Figure 7.2). The factors which make Blockchain to be used mandatorily are that the environment cannot be established in a centralized way, multiple independent authorities, without any trust factor (Abadi et al., 2018; Luu et al., 2016; Tschorsch & Scheuermann, 2016; Zheng et al., 2018).

For sharing any information, two users A and B can share Microsoft word documents with each other. A can write the information in the document and share it with B, then B can update the document and again share with A. This traditional way of sharing information does not allow both the users to update the document simultaneously. Google Docs can be used for this purpose. But the concept of Google Docs is based on the centralized environment. The major problem with a centralized system is the single point of failure. In a decentralized environment, several central coordinators are involved to handle the nodes. The system fails only when the network of nodes becomes disconnected. Thus, a more robust environment is required to handle this failure, i.e., distributed environment. In a distributed environment, no centralized coordinator is needed, thus no case of server failure.

Blockchain helps to set up such a type of distributed environment (Tschorsch & Scheuermann, 2016). In Blockchain, every user has their own copy of the information; everyone can edit his/her own copy i.e., Public Ledger and the Blockchain environment maintains the synchronization of the information.

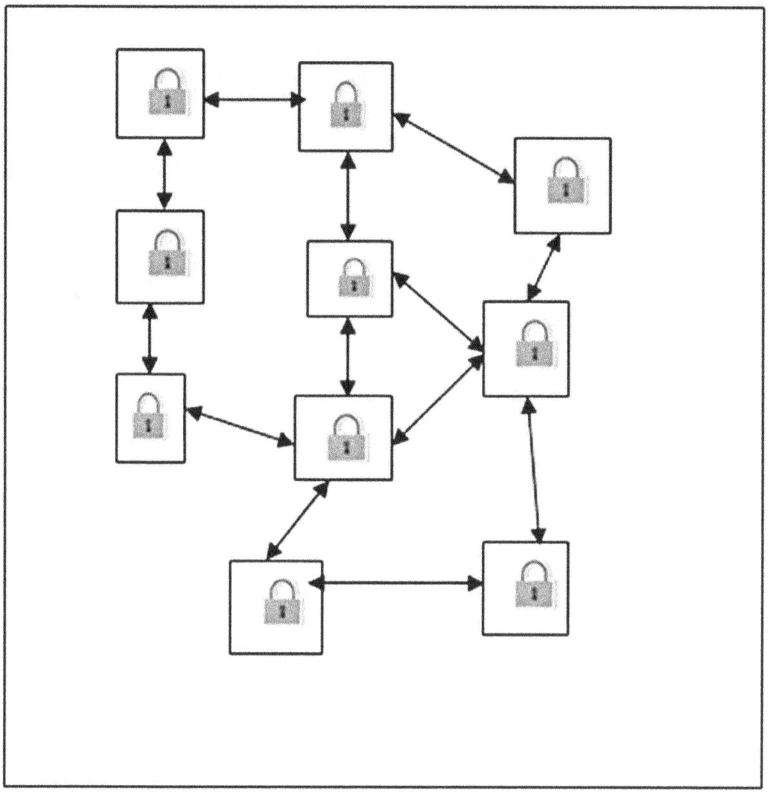

**FIGURE 7.2**
A Interconnected small square shapes with picture of locks inside.

### 7.3.1 Important Features of Blockchain

Commitment: Every operation must be validated and committed to the Blockchain (public ledger). If the operation is not valid, that must be discarded.

Consensus: Every copy with the individual users must be updated and consistent with each other.

Security: Blockchain is distributed over the individual users for their local copy. This local copy must be secure or tamperproof, nobody must be able to change or generate the false information. The security gets better with the growth of the Blockchain network.

Privacy and authenticity: The information from various users is available with every user in the Blockchain network, so it becomes mandatory to take care of the privacy and authenticity of the information.

### 7.3.2 Types of Blockchain Models

Permission-less Model: Blockchain works in open environment, a large network of users can be involved. No need to take permission to enter the network, hence anyone can join the network. Time-consuming consensus algorithms are designed to maintain such types of models. Banking using cryptocurrency is the best example for this type of working model. In this type of currency mechanism, the system must not be in the control of a single participant.

Permissioned/Private/Closed Model: Not many users can participate in this type of model; only tens to a few hundred known users participate in the closed model. A consensus algorithm used in a distributed system can be implemented in this type of model, which results in faster processing as compared to a permission-less model. This model supports more security and privacy as compared to the permission-less model.

### 7.3.3 Blockchain Benefits

Distributed: Blockchain is used in a distributed environment, with no need for a centralized entity.

Immutable: No user can change the contents of the contract.

Permission-less: Blockchain works in an open environment, hence a large number of participants can participate in the network.

Security: As the Blockchain increases, it's hard to make changes in the network, hence the Blockchain system is tamperproof.

Anonymity/privacy: The user's public key address is generated cryptographically. Nobody can understand the actual users, thus maintaining anonymity.

### 7.3.4 Technology for Secure Blockchain Architecture

Thus, Blockchain can be termed as open distributed ledger that contains the operations between the users in an efficient, verifiable, and permanent manner (Iansiti & Lakhani, 2017).

Blockchain concept is based on cryptographic secured hash function. The hash function is any hard calculation based one-way function. The popular hash functions are MD5 and SHA256 which are already proved to be cryptographically secure. Hs0 (Figure 7.3) is calculated by the user initially while entering the information and then Hs1 is calculated for the entered information. Next user, while making any editing will generate next hash value Hs2 with the current hash Hs1 and so on. Thus, whenever anyone wants to change this record of Hs1, subsequent records with hash values Hs2 and all subsequent hash values will also need to be changed and the reflected change in Hs1 will be easily noticed. Hence the cryptographic secure hash values make Blockchain tamperproof.

To calculate Hs for current information, four parameters are used i.e., sequence no, ID of the user, Timestamp (time of change of document) and the previous hash (Figure 7.3).

$$\text{Current hash} = \text{Hash}\left(\text{SNo, User's ID, Timestamp, Previous hash}\right)$$

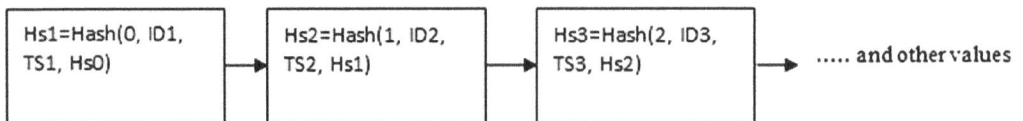

**FIGURE 7.3**
Three rectangular shapes continuing further in sequence. Every shape is depicting Hash parameters.

The basic principle of Blockchain can be used in any system where decentralized environment is used or where centralized system cannot be possible due to speed requirement, cost reduction, and security requirement. Smart contracts are used to implement a decentralized contract in a faster, cheaper, and secure manner. Smart contract stores the initial agreement code between the two parties, whenever any action takes place the initial code in the smart contract is checked and the action is executed accordingly.

For achieving above mentioned different features, cryptographic technology (hash function) is used. Hash functions take the string (arbitrary size) as input and generate output of fixed length, called message digest. Hash functions are used to make the content tamperproof. Different properties must be satisfied to have a good cryptographic hash function. First property is that the hash function must be collision free. The complexity of finding the matching hash output of n bit is $2^{n/2}$. In Blockchain architecture by using SHA256, the attacker can generate matching hash output with complexity $2^{128}$ (nearly $10^{28}$ years). Next property of hash function is information hiding; given message digest, it must be computationally hard to find the original message. Hardness of this problem depends on the size of the message digest. If X is the message and H(X) is the message digest; computation of X must be hard from the given H(X). The size of H(X) is 256 bits; the size is big enough to find X by guess. The third and last property is that if $H(X|k)$ and X is known, finding k must be computationally difficult.

Brief overview of SHA256 as Hash Function used in Blockchain Technology:

- Size of Message (M) must be the multiple of 512. Pad the message if even if it is not needed; add a single bit "1" at the end followed by k-1 0's such that l + k = 448 mod 512. Add a 64-bit block consisting of the value of l (in binary).
- Divide the complete message into blocks with length 512 bit.
- Every block is further subdivided into blocks with length 32 bit (16 sub blocks in a block).
- Compute hash values by considering $H_0$ as the initial hash value of length 256 bits: Computation of each part of M is done as $H_i = H_{i-1} + Comp_{Mi}(H_{i-1})$; (with compression function of SHA). Finally, $H_N$ is the desired digest with length 256 bits.

This architecture is based on the calculation of hash function from level 2 (the transaction level) to level 1 and then to root level. Here root level tree architecture (Figure 7.4) is based on the Merkle method (Merkle, 1987). According to this architecture (Figure 7.5) any changes made in any transaction will be reflected in the root hash and will be identified very easily; thus, changes cannot be done in any of the transaction by anyone.

In the architecture of Blockchain (Figure 7.5), the root (calculated from the set of transaction) and the previous hash are used, which guarantees the tamperproof of Blockchain. Any change in any transaction will be reflected in the root and then in the calculation of block hash and ultimately in the entire Blockchain.

Digital signatures: A digital signature is another cryptographic aspect that is used in Blockchain architecture. Digital signatures are used for content authentication, sender identity, and non-repudiation. Digital signatures cannot be reused for other documents, hence prove the identity of the user and authentication of the document.

For the implementation of all these features of digital signatures, public key cryptographic method is used. In public key cryptography two keys are used, i.e., private key ($K_{pr}$) and public key ($K_{pu}$). For digital signatures, the sender signs the document by

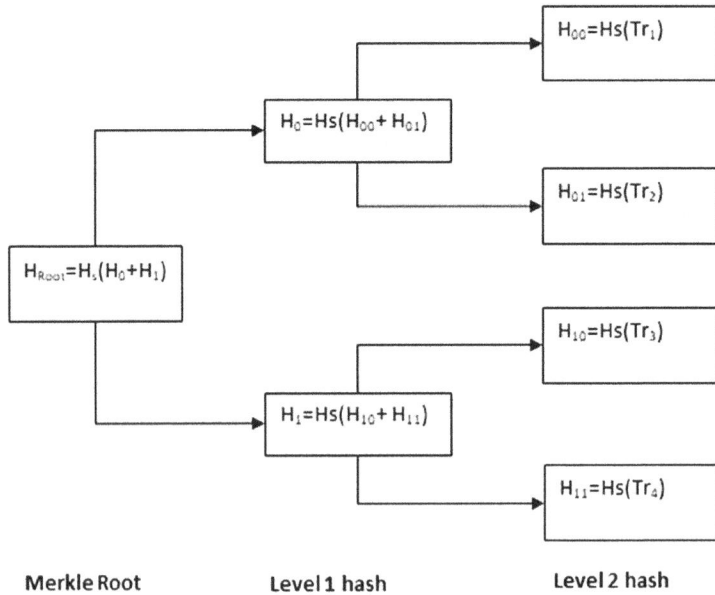

| | | $H_{00}=Hs(Tr_1)$ |
| $H_0=Hs(H_{00}+H_{01})$ | | |
| | | $H_{01}=Hs(Tr_2)$ |
| $H_{Root}=H_s(H_0+H_1)$ | | |
| | | $H_{10}=Hs(Tr_3)$ |
| $H_1=Hs(H_{10}+H_{11})$ | | |
| | | $H_{11}=Hs(Tr_4)$ |

**Merkle Root**               **Level 1 hash**               **Level 2 hash**

**FIGURE 7.4**
Horizontal tree like structure of rectangular shapes depicting the Hash values.

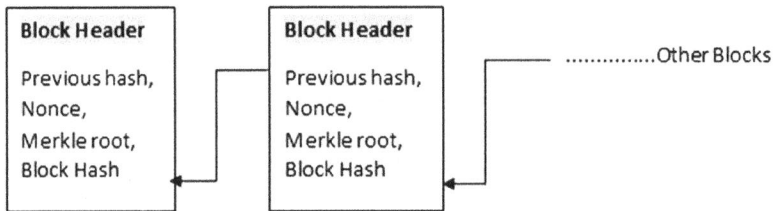

**Block Header**

Previous hash,
Nonce,
Merkle root,
Block Hash

**Block Header**

Previous hash,
Nonce,
Merkle root,
Block Hash

...............Other Blocks

**FIGURE 7.5**
Two rectangular shape continuing in sequence converging to first shape. Every rectangular shape is depicting the Block Header with its parameters.

encrypting the message by his/her private key ($K_{pr}$) which is only known to the sender. Anyone having the sender's public key can verify the message by encrypting the received message by the sender's public key. Public key algorithms are time and storage consuming, hence very costly. The complete message if signed takes plenty of resources, CPU time, and storage. For creating digital signatures only, the message digest (fixed and small length) is encrypted, resulting in less resources requirement. Hs(Msg) is the message digest of the message (Msg) to be signed, Enc is the public key method with key $K_{pr}$ and $K_{pu}$ for signing and verifying respectively (Figure 7.6).

There are many public methods available for digital signatures. In Blockchain, any of the available public key methods can be used.

The complete Blockchain architecture is given with hash and signature of the transaction involved (Figure 7.7), making the system tamper-proof and providing identity, authentication, and non-repudiation features.

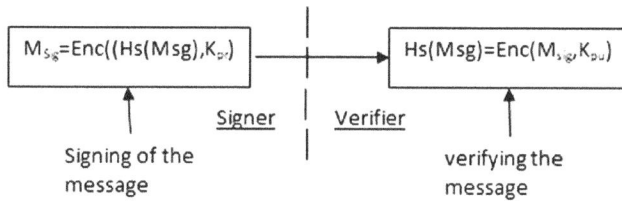

**FIGURE 7.6**
Two rectangular shapes depicting the formula for signer and verifier divided by a dotted line.

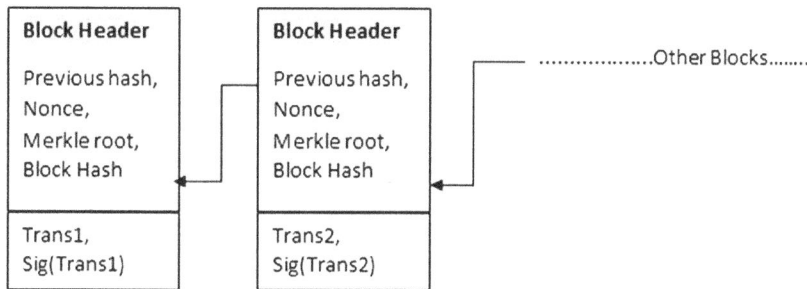

**FIGURE 7.7**
Two rectangular shapes continuing in sequence converging to first shape. Every rectangular shape is depicting the Block Header with its parameters and Transaction signatures.

## 7.4 Integration of Blockchain and IoT

Blockchain plays a vital role in Internet of Things by increasing security (Bahga & Madisetti, 2016; Christidis & Devetsikiotis, 2016; Conoscenti et al., 2016; Kshetri, 2017; Nartey et al., 2021; Polyzos & Fotiou, 2017; Reyna et al., 2018, Skwarek, 2017; Truong et al., 2019; Uddin et al., 2021; Yeow et al., 2018; Zhang et al., 2019). In IoT-like systems, Blockchain provides trust and transparency. Blockchain helps in establishing the system where a centralized system (client server model) cannot be feasible. For example, in IoT systems, critical information like sensitive location and weather information can be updated using Blockchain. This updating is done by all involved parties with some terms of agreements. There are many other advancements for which the researchers are continuously working for Blockchain integrated IoT systems (Atlam et al., 2020; Fernandez-Carames & Fraga-Lamas, 2018; Fukuda & Omote, 2021; Glaser, 2017; Korpela et al., 2017; Wang et al., 2019; Yavari et al., 2020; Yue et al., 2020).

### 7.4.1 Advantages of Using Blockchain IoT

**Avoiding Single Point Failure**: In centralized systems, single point failure is the big fear. Blockchain technology distributes the computation and storage over millions of devices. Thus, the failure of any devices does not mean the failure of the complete system, resulting in more robust and reliable system.

**Building Trust**: Blockchain helps in building the trust between different parties. In Blockchain system, cryptographic algorithms are used, which ensures the security and confidentiality of the system. Thus, unverifiable party can join the network without being verified by the other parties. Blockchain does this work for all the parties by storing the information and exchanges permanently in the network. Also, the man in the middle attack is not possible in this kind of system as no single communication can be intercepted. The complete history is recorded in the system, which is completely tamperproof.

**Reducing the Cost**: In Blockchain, IoT gateways are not required, resulting in less processing. The cost of installing the expensive server and then their maintenance in centralized system can be avoided using Blockchain. All the processing is done on peer-to-peer basis. Thus, Blockchain helps in providing a significant reduction in cost.

**Improvement in Data Storage and Exchange Rate**: In the peer-to-peer system, storage requirement is also distributed over all the devices (millions in count). Besides cost reduction, Blockchain helps to accelerate data exchange. As IoT gateways or middle protocols are not required in Blockchain, time-consuming processing done in between need not to be done, hence overall speed increases.

**Scaled Security**: Centralized system can establish the security requirements in an efficient way when the number of devices in the network is small. Security features attainment becomes complex (almost impossible) as the system grows. Blockchain is capable to fulfill this demand in an efficient manner.

In the current IoT system, all the devices must be on the internet for their connection, identification, and authentication. Cloud servers are used for all these connection features. This cloud framework can handle only a small IoT system, as the system grows the cost increases exponentially. Also, there is still the requirement of a single platform for connecting all the diverse devices. Thus, the centralized system cannot serve the current requirement of IoT systems, where millions and billions of devices are supposed to be connected.

## 7.5 Analysis of Various IoT Applications with Blockchain

Blockchain provides the secure way for industrial applications which are to be implemented on a distributed environment (Ducrée, 2020; Javaid et al., 2021; Stallone et al., 2021). Following are some of the applications for which Blockchain provide the suitable framework.

### 7.5.1 Supply Chain

In Supply Chain, for tracking the movement of the products, generally expensive items, need robust mechanism like Blockchain. It manages the link between multiple manufacturers, shippers, distributor, and retailer. One need to check the movement of the items from one user to other. Blockchain in permissioned environment is used to implement this. None of the receiver can deny that they have not received the product. Token is generated in Blockchain at the starting location of the product that is considered to be the trusted entity. In parallel movement of product from one location to other, the token also transfers.

Tokens are added in the Blockchain at every location which gives the updated information of the product's location. Thus, keeping the token location information in Blockchain the supply chain track can be managed in a secure way.

### 7.5.2 Ride Hiring System

In the Ride Hiring System, currently, the taxis are hired on a centralized server, e.g., Uber, Lyft, and many others. The complete database is handled by the service provider. The user has a separate account for different providers and the users choose one to depend on the service provided by the provider. Service provider is responsible for maintaining all the interaction of the user on that account. This type of maintenance is costly and time consuming. At the worst, if the service provider is closing the application all the transactions associated with the devices under the provider will be lost, even one cannot find that device.

On the other hand, in Blockchain-based system only one account is required by the user. Single account can interact with each transaction from different service. The devices and the users are directly connected. The rules are framed and written as smart contract; the users can use the devices based on the smart contracts.

### 7.5.3 Solar Energy

In a community system to manage economy, every community share one or the other resources. In a community, solar energy is used for power consumption. The device is used to record the production. This record can be handled by Blockchain and anyone requiring the solar energy can directly contact to the device and use the energy based on rule framed in the smart contracts. The rule can be framed depending upon the total transaction and maintenance history of the user. In this way the device can be used to purchase solar power and users can share the maintenance of the device.

### 7.5.4 Printing Machine

Ownership can be shared in a printing machine, e.g., 3D printers, crafting machine, etc. The agreement on smart contract is used to share the ownership of the machines for crafting any pattern. The machine maintenance is shared by the users of any community, corporate buildings etc. The framed rules can be based on the maintenance of the machine and the owner's agreement.

### 7.5.5 Home Space

Users can have the requirement to access a house on a rental basis. Depending on the framed rule by the owner, the user can access the house, otherwise the house will not be accessible. The users who may have left the house in a bad condition with some damage will be blocked from accessing the home. A smart contract can be made on the basis of whether the user has leased out the house more than twice with good reviews.

### 7.5.6 Waste Management

Every user in the community must perform at least four maintenances in a year. The Blockchain helps in maintaining that record. Every time the device itself will call the maintenance man whenever required and the users have to manage the finance. The smart contract will record the user's transactions. The user not satisfying the framed rule can be given alert or their wastes disposal can be stopped depending upon the agreement.

### 7.5.7 Agriculture System

The users in the agriculture field can share the water pump using Blockchain technology. The users who have not participated in maintenance of the water pump will be blocked from using the service. Also, if the user fails to satisfy any agreement of the contract, the pump will not be unlocked for that user.

### 7.5.8 Healthcare System

Many wearable devices are in trend, e.g., heartrate monitors, fit bands, smartwatches, and others. Healthcare smart devices might communicate with hospital management for further action in any urgency. For example, the heartrate monitoring device may call the ambulance in case the device finds the heartbeat at abnormal rate. In this type of system if the wearer is having some good designation in a company and his opponent does not want him to be present in some important meeting, the opponent may tamper with the data of the wearable device. Blockchain can help to keep the privacy and secrecy of the health data.

### 7.5.9 Miscellaneous Applications

In **Smart Home Appliances**: Blockchain helps home appliances to behave as independent agents. Many appliances like washing machines, vacuum cleaners, and dishwashers make their priority to work to save the power requirement.

**Smart Vending Machine**: With the help of Blockchain, a vending machine can work as an independent entity. The vending machine can monitor its stock and make the report, besides this with the help of Blockchain technology the machine can send the demand to the distributors depending on the customer purchasing history and pay for the same.

**Smart Vehicle**: Vehicle can autonomously check its own maintenance requirements, call the engineer, and pay for it by using Blockchain technology.

In the Voting System, today's scenario can be challenging for Electronic Voting Machines (EVMs). With the help of Blockchain technology, the voting system can be made tamperproof. Every vote can be treated as a block in Blockchain, resulting in a tamperproof voting system. This system will also support transparency in the process.

In property ownership, maintaining the property record the current system of official records is complex with lots of burdens. Also, this system is very inefficient because of the manual entries. The property disputes become difficult to solve. Thus, the current system is costly, time-consuming, and full of human errors. With the use of Blockchain technology, there is no need to scan the official file in the records. Also, the owners will be having more trust in the ownership deal.

### 7.5.10 Application Implementation

For generating decentralized applications, Hyperledger Fabric, Ethereum, and Corda are the popular Blockchain platforms. Various tools for Blockchain development are used, like

{"customer":"Happy","index":4,"location":"Faridabad","message":"A block is
MINED","owner":"Sai","previous_hash":"43f64e1d39583b9241edafc57e784ca5ff6e2515c166a385001567e5bd40da59","proof":"147942","tim
estamp":"2021-06-16 14:49:44.556710"}

**FIGURE 7.8**
Block value starting with curly brackets containing parameters with their values for the mined block.

{index: 1, timestamp: '12:10:00', proof: 0, previous_hash: '0', owner: 'karan', customer: 'xyz', location: 'delhi'}

{index: 4, timestamp: '2021-06-16 14:49:44.556710', proof: 147942, previous_hash: '43f64e1d39583b9241edafc57e784ca5ff6e2515c166a385001567e5bd40da59', owner: 'Sai', customer: 'Happy',
location: 'Faridabad'}

{index: 5, timestamp: '2021-06-16 15:05:20.983231', proof: 183854, previous_hash: '65428491f844910f ece455f429c76f91b15a8a52c6ef36644e79951f3bd8055a7', owner: 'Sudhakar', customer:
'Harpreet', location: 'Delhi'}

{index: 6, timestamp: '2021-06-16 15:16:48.765122', proof: 26255, previous_hash: '21ab00f6e818109659c6bd2eb28b61f78eefa5403987a888e271d9624b47b', owner: 'Krishan', customer: 'Kiran',
location: 'Rajasthan'}

{index: 7, timestamp: '2021-06-16 15:52:53.080750', proof: 111896, previous_hash: 'b816f944366f288aa94565844600250451c59568f5e62802705673a23348ad25', owner: 'Krishan', customer: 'Kiran',
location: 'Rajasthan'}

**FIGURE 7.9**
Block value starting with curly brackets containing Blockchain parameters and values of different mined block.

Solidity (object-oriented programming language), Geth (uses Go programming language), Remix, Mist, and many others. Here the details are omitted for the mentioned tools. For a simple illustration of the application, Blockchain output (Figures 7.8, 7.9) is presented using Python:

Few Python libraries are used; like JSON for returning Blockchain, datetime for time-stamp, and hashlib for generating the hash.

## 7.5.11 Analysis of Blockchain Integrated IoT Applications

Researchers are continuously working in the area of Blockchain technology since its inception as described above. Healthcare (Antwi et al., 2021), car-sharing (Saurabh et al., 2021), power grids (Foti & Vavalis, 2021) are few applications implemented recently. Following (Table 7.1) is the analysis of various IoT applications with their smart contract rules:

**TABLE 7.1**

Analysis of Blockchain Integrated IoT Applications

| Application | Utility | Smart Contract Rule |
|---|---|---|
| Supply Chain | Manufacturers, shippers, distributor, and retailers | Token location information track |
| Ride Hiring | Owners and costumers | Traffic rule violation, vehicle maintenance etc. |
| Solar Energy | Society, corporates | Device maintenance by users |
| Printing Machine | Corporates, community | Maintenance by users and owner's agreement |
| Home Space | Owner and users | User's previous reviews and owner's agreement |
| Waste Management | Municipal corporation and users | Maintenance Finance by users |
| Agriculture | Irrigation system | Pump maintenance by users |
| Healthcare | Patients and medical data security | Record access control |

## 7.6 Conclusion and Summary

Currently, centralized communication servers are handling all critical tasks; but with the exponential growth of the IoT systems one has to shift from centralized to distributed environment with Blockchain which guarantees the security for the system in this unreliable environment. Blockchain provides the solution to many IoT security challenges. With Blockchain, tamper-proof ledger can be seen and authenticated to avoid any kind of privacy and authentication issues. Many applications are presented with their analysis to showcase the vast exposure of Blockchain in IoT industry. In near future everything is to be evolved as IoT and to take the challenge of IoT security Blockchain is the only solution. Budding researchers may make their career in this field to come up with the solution of various IoT applications with Blockchain technology.

## References

Abadi, F.A., Ellul, J. and Azzopardi, G. (2018, July 30–August 3). The blockchain of things, beyond bitcoin: a systematic review [Conference session]. *Ist International Workshop on Blockchain for the Internet of Things*, Halifax. doi:10.1109/Cybermatics_2018.2018.00278.

Alqassem, I. and Svetinovic, D. (2014, December 9–12). A taxonomy of security and privacy requirements for the Internet of Things (IoT) [Conference session]. *International Conference on Industrial Engineering and Engineering Management, Selangor.* doi: 10.1109/IEEM.2014.7058837.

Antwi, M., Adnane, A., Ahmad, F., Hussain, R., Rehman, M.H. and Kerrache, C.A. (2021). The case of hyperledger fabric as a blockchain solution for healthcare applications. *Blockchain: Research and Applications*, 100012, ISSN 2096-7209. doi: 10.1016/j.bcra.2021.100012.

Atlam, H.F., Azad, M.A., Alzahrani, A.G. and Wills, G. (2020). A review of blockchain in Internet of Things and AI. *Big Data and Cognitive Computing*, 4(28). doi: 10.3390/bdcc4040028.

Atzori, L., Iera, A. and Morabito, G. (2010). The Internet of Things: A survey. *Computer Networks*, 54(15), 2787–2805. doi: 10.1016/j.comnet.2010.05.010.

Bahga, A. and Madisetti, V.K. (2016). Blockchain platform for industrial Internet of Things. *Journal of Software Engineering and Applications*, 9(10), 533–546. doi: 10.4236/jsea.2016.910036.

Bi, Z., Xu, L.D. and Wang, C. (2014). Internet of Things for enterprise systems of modern manufacturing, *IEEE Transactions on Industrial Informatics*, 10(2), 1537–1546. doi: 10.1109/TII.2014.2300338.

Christidis, K. and Devetsikiotis, M. (2016). Blockchains and smart contracts for the Internet of Things. *IEEE Access*, 4, 2292–2303. doi: 10.1109/ACCESS.2016.2566339.

Conoscenti, M., Vetro, A. and de Martin, J.C.(2016, November 29–December 2). Blockchain for the Internet of Things: A Systematic Literature Review" [Conference session]. *IEEE/ACS 13th International Conference of Computer Systems and Applications (AICCSA)*, Agadir. doi:10.1109/AICCSA.2016.7945805.

Ducrée, J. (2020). Research – A blockchain of knowledge? *Blockchain: Research and Applications*, 1(2), 100005, ISSN 2096-7209.doi:10.1016/j.bcra.2020.100005.

Evans, D. (2011). The Internet of Things: How the next evolution of the internet is changing everything. www.cisco.com/c/dam/en_us/about/ac79/docs/innov/IoT_IBSG_0411FINAL.pdf

Fernandez-Carames, T.M. and Fraga-Lamas, P. (2018). A Review on the use of Blockchain for the Internet of Things. *IEEE Access*, 6, 32979–33001. doi: 10.1109/ACCESS.2018.2842685.

Foti, M. and Vavalis, M. (2021). What blockchain can do for power grids? *Blockchain: Research and Applications*, 2(1), 100008, ISSN 2096-7209. doi: 10.1016/j.bcra.2021.100008.

Fukuda, T. and Omote, K. (2021). Efficient Blockchain-based IoT Firmware Update Considering Distribution Incentives [Conference session]. *IEEE Conference on Dependable and Secure Computing (DSC)*. doi: 10.1109/DSC49826.2021.9346265.

Glaser, F. (2017, January 4–7). Pervasive Decentralisation of Digital infrastructures: a framework for blockchain enabled system and use case analysis [Conference session]. *50th Hawaii International Conference on System Sciences*, Waikoloa Village, HI. doi: 10.24251/HICSS.2017.186.

Iansiti, M. and Lakhani, K.R. (2017). The truth about blockchain. *Harvard Business Review*, 95(1), 18–127. https://hbr.org/2017/01/the-truth-about-blockchain

Javaid, M., Haleem, A., Singh, R.P., Khan, S. and Suman, R. (2021). Blockchain technology applications for Industry 4.0: A literature-based review. *Blockchain: Research and Applications*, 100027, ISSN 2096-7209. doi: 10.1016/j.bcra.2021.100027.

Korpela, K., Hallikas, J. and Dahlberg, T. (2017, January 4–7). Digital supply chain transformation toward blockchain integration [Conference session]. *50th Hawaii International Conference on System Sciences*, Waikoloa Village, HI. doi: 10.24251/HICSS.2017.506.

Kshetri, N. (2017). Can blockchain strengthen the Internet of Things. *IT Professional*, 19(4), 68–72. doi: 10.1109/MITP.2017.3051335.

Luu, L., Chu, D.H., Olickel, H., Saxena, P. and Hobor, A. (2016, October 24–28). Making smart contracts smarter. [Conference session]. *ACM SIGSAC Conference on Computer and Communications Security*, Wien. doi: 10.1145/2976749.2978309.

Merkle, R.C. (1987, August 16–20). A digital signature based on a conventional encryption function. [Conference session]. *7th Conference on Advances in Cryptology (CRYPTO)*, Santa Barbara, CA. doi: 10.1007/3-540-48184-2_32.

Miorandi, D., Sicari, S., de Pellegrini, F. and Chlamtac, I. (2012). Internet of Things: Vision, applications and research challenges. *Ad Hoc Networks*, 10(7), 1497–1516. doi: 10.1016/j.adhoc.2012.02.016

Nakamoto, S. (2008). Bitcoin: A peer-to-peer electronic cash system. https://bitcoin.org/bitcoin.pdf

Nartey, C., Tchao, E.T., Gadze, J.D., Keelson, E., Klogo, G.S., Kommey, B. and Diawuo, K. (2021). On Blockchain and IoT Integration Platforms: Current Implementation Challenges and Future Perspectives. *Wireless Communications and Mobile Computing, 2021*, Article ID 6672482. doi: 10.1155/2021/6672482.

O'Halloran, D., Kvochko, E., Daugherty, P. and Reilly, M. (2015). Industrial Internet of Things: Unleashing the potential of connected products and services. http://www3.weforum.org/docs/WEFUSA_IndustrialInternet_Report2015.pdf

Pereira, T.S.M., Barreto, L. and Amaral, A. (2017). Network and information security challenges within Industry 4.0 paradigm. *Procedia Manufacturing*, 13, 1253–1260. doi: 10.1016/j.promfg.2017.09.047

Polyzos, G.C. and Fotiou, N. (2017, August 4–6). Blockchain-assisted information distribution for the Internet of Things [Conference session]. *IEEE International Conference on Information Reuse and Integration*, San Diego, CA, 75–78. doi: 10.1109/IRI.2017.83.

Reyna, A., Martín, C., Chen, J., Soler, E. and Díaz, M. (2018). On blockchain and its integration with IoT. challenges and opportunities. *Future Generation Computer Systems*, 88, 173–190. https://www.sciencedirect.com/journal/future-generation-computer-systems/issues

Sadeghi, A.R., Wachsmann, C. and Waidner, M. (2015, June 7–11). Security and privacy challenges in industrial Internet of Things [Conference session]. *52nd Annual Design Automation Conference*, San Francisco, CA. doi: 10.1145/2744769.2747942.

Saurabh, N., Rubia, C., Palanisamy, A., Koulouzis, S., Sefidanoski, M., Chakravorty, A., Zhao, Z., Karadimce, A. and Prodan, R. (2021). The ARTICONF approach to decentralized car-sharing. *Blockchain: Research and Applications*, 100013, ISSN 2096-7209. doi: 10.1016/j.bcra.2021.100013.

Schneider, S. (2017). The industrial Internet of Things (IIOT): Applications and taxonomy. In Geng, H..(Ed.), *Internet of Things & Data Analytics Handbook* (pp. 41–81). Auflage, John Wiley & Sons, Hoboken, NJ. doi: 10.1002/9781119173601.ch3

Skwarek, V. (2017). Blockchains as security-enabler for industrial IoT-applications. *Asia Pacific Journal of Innovation and Entrepreneurship*, 11(3), 301–311. doi: 10.1108/APJIE-12-2017-035.

Stallone, V., Wetzels, M. and Klaas, M. (2021). Applications of Blockchain Technology in marketing systematic review of marketing technology companies. *Blockchain: Research and Applications*, 100023, ISSN 2096-7209. doi: 10.1016/j.bcra.2021.100023.

Truong, H., Almeida, M., Karame, G. and Soriente C. (2019). Towards Secure and Decentralized Sharing of IoT Data [Conference session]. *IEEE International Conference on Blockchain (Blockchain)*, Atlanta, GA, USA. doi: 10.1109/Blockchain.2019.00031.

Tschorsch, F. and Scheuermann, B. (2016). Bitcoin and beyond: A technical survey on decentralized digital currencies. *IEEE Communications Surveys & Tutorials*, 18(3), 2084–2123. doi: 10.1109/COMST.2016.2535718

Uddin, M.A., Stranieri, A., Gondal, I. and Balasubramanian, V. (2021). A survey on the adoption of blockchain in IoT: Challenges and solutions. *Blockchain: Research and Applications*, 100006, ISSN 2096-7209. doi: 10.1016/j.bcra.2021.100006.

Wang, G., Shi Z., Nixon M. and Han S. (2019). ChainSplitter: Towards Blockchain-Based Industrial IoT Architecture for Supporting Hierarchical Storage [Conference session]. *IEEE International Conference on Blockchain (Blockchain)*, Atlanta, GA, USA. doi: 10.1109/Blockchain.2019.00030.

Wollschlaeger, M., Sauter, T. and Jasperneite, J. (2017). The future of industrial communication, Automation networks in the era of the Internet of Things and industry 4.0. *IEEE Industrial Electronics Magazine*, 11(1), 17–27. doi: 10.1109/MIE.2017.2649104.

Yavari, M., Safkhani, M., Kumari, S., Kumar, S. and Chen, C.M. (2020). An Improved Blockchain-Based Authentication Protocol for IoT Network Management. *Security and Communication Networks*, 2020, ID 8836214. doi: 10.1155/2020/8836214.

Yeow, K., Gani, A., Ahmad, R.W., Rodrigues, J.J.P.C. and Ko, K. (2018). Decentralized consensus for edge centric Internet of Things: A review, taxonomy, and research issues. *IEEE Access*, 6, 1513–1524. doi: 10.1109/ACCESS.2017.2779263.

Yue, Wu, Liangtu, S, Lei, L., Jincheng, L., Xuefei, L. and Linli, Z. (2020). Consensus mechanism of IoT based on blockchain technology. *Shock and Vibration*. doi:10.1155/2020/8846429.

Zhang, Y., Kasahara, S., Shen, Y., Jiang, X. and Wan, J. (2019). Smart contract-based access control for the Internet of Things. *IEEE Internet of Things Journal*, 6(2), 1594–1605. doi: 10.1109/JIOT.2018.2847705.

Zheng, Z., Xie, S., Dai, H., Chen, X. and Wang, H. (2018). Blockchain challenges and opportunities: A survey. *International Journal of Web and Grid Services*, 14(4), 352–375. doi: 10.1504/IJWGS.2018.095647.

# 8

## Blockchain: A New Power-Driven Technology for Smart Cities

**Shyamal Srivastava**

*VIT University, Chennai, India*

**R. Girija**

*VIT University, Chennai, India*

## CONTENTS

## 8.1  Introduction: Background and Driving Forces

Since its introduction to the world by Satoshi Nakamoto, Blockchains have found numerous applications across multiple domains apart from serving as a distributed ledger for cryptocurrencies. From healthcare to supply chain management, Blockchain technology has disrupted the way these fields have traditionally operated. Blockchain is basically a shared and distributed record or ledger to which new entries (or transactions as they are called in reference to Blockchain) can be added only through a unanimous decision by all the participants and any existing entries or records cannot be changed [1–3]. Therefore, its ideal to serve as a platform where data integrity and authenticity are of paramount

DOI: 10.1201/9781003206088-8

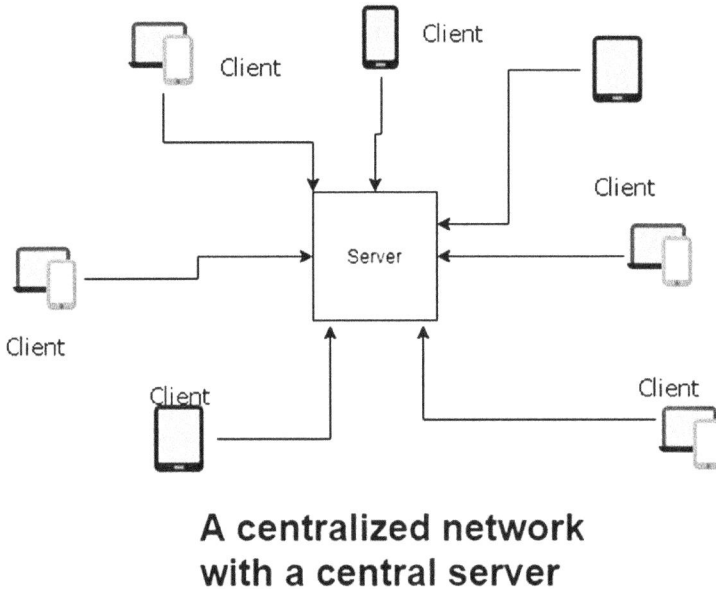

**A centralized network
with a central server**

**FIGURE 8.1**
Topology of a centralized network.

importance. Blockchain also gives real ownership to each participant on the network. This can provide a sense of inclusiveness and community building among the residents of Smart Cities. Any system implemented using a Blockchain is safer and more transparent. Thus, it is an ideal platform for storing public records. Figures 8.1 and 8.2 show the topological structure of a centralized and decentralized network respectively.

As new technologies are emerging, they are changing every aspect of our lives. A smart city is a city that leverages the latest technologies to improve the lives of its citizens. Smart cities make use of technologies like the Internet of Things (IoT) and machine learning to do things more efficiently. Smart cities utilize a mix of information collection, handling, and distribution along with networking and computer science to improve upon traditional ways in which various services like education, traffic control, energy distribution, etc. are handled by the municipalities and local government institutions [4–7].

Blockchain is another such technology that can revolutionize the way that traditional cities have operated. When used alongside other technologies like IoT and machine learning, Blockchain can act as a very powerful tool which can be utilized by Smart Cities. In this chapter, we discuss various use cases of Blockchain and how they can help improve Smart Cities [8].

## 8.2 Solar Energy Smart Grid Powered with Blockchain

Smart Grid can be defined as an electrical grid with automation, communication and IT frameworks that collects and analyses data regarding power flow from the source to points

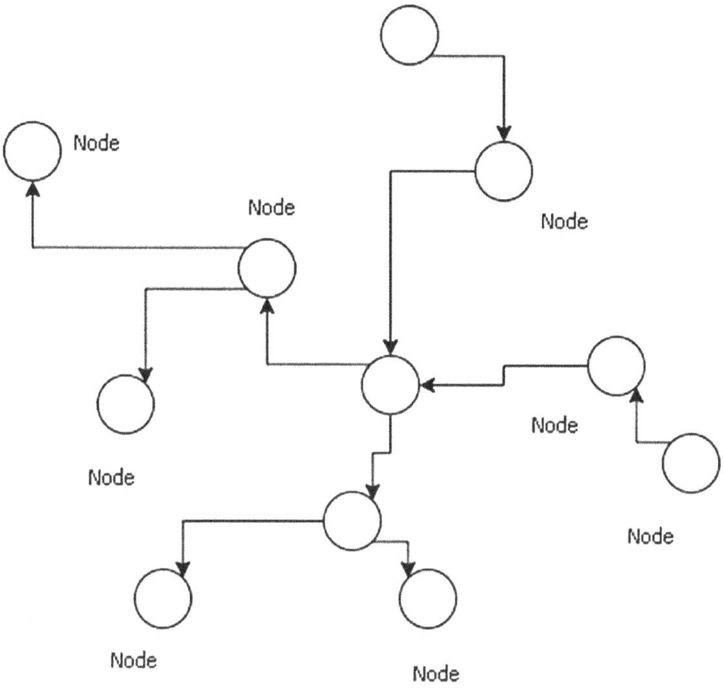

**FIGURE 8.2**
Topology of a decentralized network.

of utilization and control the flow of electricity according to data collected by sensors present at various locations throughout the smart grid. Smart grid solutions help to monitor, measure and control power flows in real time that can contribute to identification of losses and thereby appropriate actions can be taken to mitigate the power losses.

Imagine a neighborhood where every house has solar panels on its roof and the neighbors can trade solar energy generated by their solar panels among each other. A big challenge with solar energy is storing the energy as batteries available today are not very cost effective. This could lead to a loss of surplus energy generated. A smart grid that allows trading of energy among peers could minimize this loss. Energy surplus can be sold while energy can be bought from others in case of shortfall. If the smart grid is backed by a Blockchain, the transactions can be easily done in a transparent and automated manner using crypto tokens. Blockchain provides a reliable way to sell and buy energy among the participants of the smart grid. A computer or an embedded system can be used as nodes for the network. Each participant will have a node which will act as a router for energy and also double up as a node for the network. Figure 8.3 shows a layout for such a smart grid. Many solutions for energy sharing have already been developed which utilize Blockchain technology. With advancements made in science and technology, solar panels are getting more efficient every day. It is possible in the near future that we can develop neighborhoods having a Blockchain powered solar smart grid without any reliance on other energy sources [9–11].

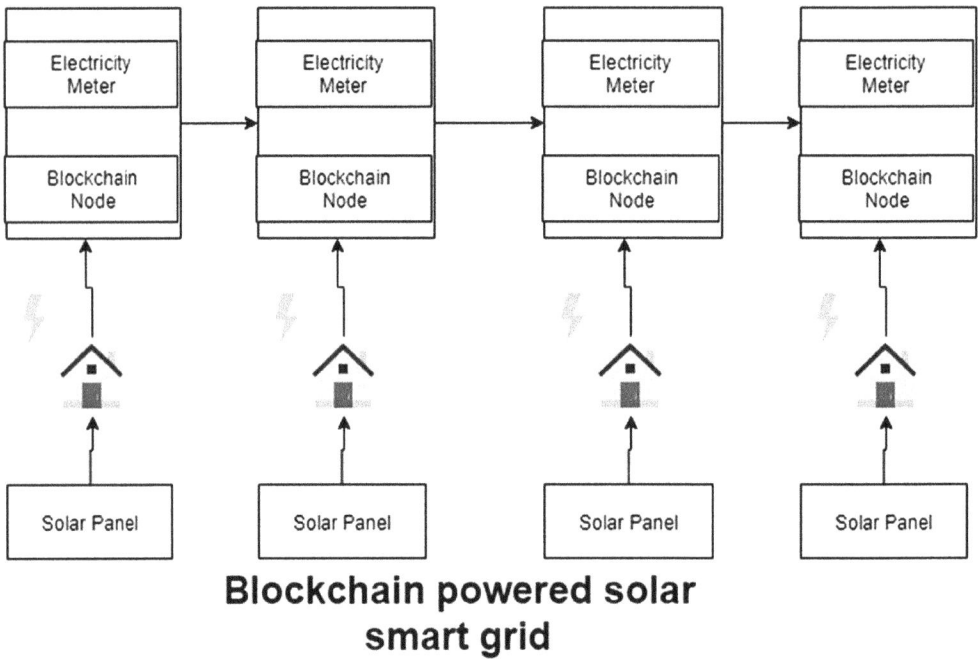

## Blockchain powered solar smart grid

**FIGURE 8.3**
Blockchain powered smart grid.

## 8.3 Feedback and Opinion Gathering with Blockchain

For local authorities, public feedback can be very important in matters of developing new projects in the city or gathering data to help formulate policies. While traditional polls can be used for this purpose, they are generally not effective in engaging people as they lack much incentive to participants. Participants would not answer honestly or might give random answers if they are not interested in the feedback survey or poll. This issue can be addressed by moving the feedback survey or poll to a Blockchain-based platform where users whose answers end up to be the most common answers get automatically rewarded with crypto tokens. This would motivate participants to be thorough and honest. Conducting an opinion poll also costs money. If the sample space for an opinion poll is not wide enough, it is impossible to get accurate results. There is a phenomenon known as 'wisdom of crowds' at work in such cases. If the polls are moved to a Blockchain, the cost can be cut down drastically and the money saved could be used as a bonus for people who voted for the most popular choice in the poll. This will encourage people to participate and give their honest opinions resulting in far more accurate and wider polls. This can immensely help local authorities in collecting information and opinions of the residents of the smart city [12–15].

## 8.4 Conducting Elections and Referendums

Conducting elections during the COVID-19 pandemic is a very challenging task with many elections becoming super spreader events. A digital form of voting is the need of the day for conducting elections remotely. In US presidential elections, voters were given a choice of mail-in voting but it was a highly controversial move as many questioned the authenticity of such votes. The risk involved with conducting online voting is very high as hackers can potentially hack the results of the election. Using Blockchain reduces this risk to almost negligible as it is almost impossible to hack a robust application running on a Blockchain. An electronic approach to use Blockchain innovation to get an individual's vote and not permit any political decision official or political individual to change a vote can revolutionize how elections are conducted. From national level elections to municipal elections and referendums for city residents can be conducted in this way. This will not only make the voting process easier and more accessible but also cut down the cost.

While a traditional network might not be suitable for conducting important elections due to security concerns and risks of the election process getting hacked, it is almost impossible to hack a Blockchain network as hackers must have control over at least half of all the nodes on the network at the same time. This is very unlikely to happen.

## 8.5 Citizen Rating Based on Environmental Awareness

Blockchain can be used to give environmental awareness ratings to citizens to promote an Eco-friendly lifestyle among them. Various information like energy usage, water usage and use of public transportation can be tracked through various IoT-based sensors that communicate their observations to the Blockchain where this data can be used to calculate an environmental awareness index for each citizen. People who have a good index can be rewarded in form of tax write off or discounts on their energy bill. An extra can be levied for users who have a larger carbon footprint. This concept could be expanded to reward users when they give waste for recycling in waste bins fitted with sensors. Some amount of cryptocurrency or tokens can be paid to them for their efforts to recycle wastes. Citizens can be motivated to be more mindful of their environment if they are rewarded for positive changes in their lifestyle. Similar programs have been introduced in certain countries already. For example, in China commuters can get discounts on their train tickets if they give their used recyclable plastic bottles in exchange. Small programs like these can go a long way in creating Smart Cities which are carbon neutral.

Use of Blockchain will ensure that the system is tamper proof and transparent. The entire history of a citizen regarding their impact on the environment can be stored and accessed as and when required. This can also be made available to third parties like businesses or stores willing to give additional discounts or benefits to citizens with outstanding history of environmental awareness. Similarly, the price for fossil fuels can be dynamically adjusted for each individual. For example, the first 5 gallons of petrol per month can be bought at a base price but after that additional cess can be levied on any excess fuel consumption. This will encourage citizens to move toward clean and renewable sources of energy.

## 8.6 Secure Storage of Private and Sensitive Data of Citizens

When storing sensitive data the following criteria's must be kept in mind:

- The data must only be accessible by authorized parties
- Multiple copies must be stored to prevent any loss of data
- Modification or deletion of the data must only be done by authorized nodes

Permissioned Blockchains can fulfill all of these criterias and therefore are suitable for this use case. Local governments can utilize permissioned Blockchains for securing sensitive data like tax returns and social security numbers of their citizens. Figure 8.4 shows how a permissioned Blockchain works. Blockchain is protected with strong encryption to ensure that only nodes with necessary permissions can access or change data. Also, unlike traditional servers, Blockchain does not have a single point of failure. Data stored on Blockchain is extremely persistent as copies are stored across various nodes. So in case one or more nodes fail, the data will still be secure and authentic. The data present on the blocks of the Blockchain as well as the operations performed by the different members on the Blockchain can be controlled relying on how the Blockchain is arranged and how it is relied upon to satisfy the desired purpose. Data can be stored in other decentralized systems which have been present from long before the introduction of Blockchain. There are various decentralized information base frameworks on which to assemble apps that take into consideration this component like distributed hash trees, downpour, and interplanetary file systems. Public and private Blockchains are the two most basic varieties of Blockchain. They are utilized vigorously among the different open networks like Ethereum and Cardano and private organizations. A third-class, permissioned Blockchain has additionally acquired footing. Permissioned Blockchains can ensure that data is viewed only by authorized nodes thus minimizing the possibility of a data leak. Alternatively, for some applications, private Blockchains can be used. However, not everyone can access a private Blockchain making its use cases limited.

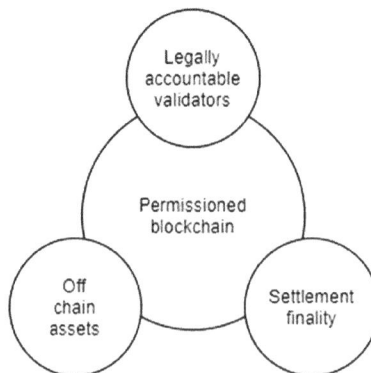

**FIGURE 8.4**
The workings of a permissioned Blockchain.

## 8.7 Generating New Streams of Income for Local Administrative and Municipal Bodies

In smart urban communities, all sensors and gadgets connected with an IoT network have the ability to get and send information and execute orders, empowering distant checking and dynamic, versatile control capacities. Smart lighting is a famous model: by gaining information from streetlights and connecting it with scheduled events, ecological conditions, or vehicle travel, we can switch and diminish lights precisely where and when required, in this manner lessening energy utilization, streamlining costs, improving nature of administration and boosting resident fulfillment. Be that as it may, imagine a scenario where gadget-related information could be adapted and transformed into an income opportunity. On account of Blockchain, information from streetlamps, parking garages, squander receptacles, natural sensors, and other metropolitan articles can be changed into tradable tokens. Think about the data a leaving sensor produces about the vehicle parcel being empty or occupied. It can be turned into a token and exchanged; stopping administrators can purchase these "data tokens" to plan and offer their own brilliant stopping administrations. These can fuel private organizations, while urban areas move their stopping sensor venture toward income, simultaneously as profiting by lower traffic and contamination, and better quality of life.

Municipalities can create a stream of passive and periodic income without requiring a lot of work if they can identify useful data which can be utilized by other parties willing to pay for it. This extra income can recover all the costs of setting up Blockchain-based solutions for the smart city in a short period of time.

## 8.8 Blockchain-Based Universal Identification

Data stored on Blockchain is extremely secure as it is protected by strong encryption and persistent as multiple copies are stored across the network. Identification documents are sensitive information which can be put to the wrong use in case they fall into the wrong hands. Governments across the world are coming up with unique identification for their citizens. In India, the Aadhar card serves the purpose of being a single identification document that can be used in most places.

However, a grave concern for governments is the forgery of identification documents and identity theft. While authorities can store the digital records in a server, counterfeit identification documents can be added to the central server if the security measures are bypassed by a hacker. Nowadays, it is common to hear servers of government bodies or big corporate organizations being compromised by cyber-attacks. In the case of a Blockchain, however, a hacker needs to gain access to at least 50 percent of all the nodes present on a network simultaneously in order to make any changes without getting consensus and even that might lead to a fork in the Blockchain leaving all their efforts useless. Therefore, Blockchain can serve as an ideal platform for storing and issuing new identification documents. Provisions can be made to decide which users have the authority to issue new nodes using simple access-controlled codes in smart contracts making the entire process very simple. Land ownership records and property titles are documents and records which prove ownership of the concerning property. The entire system can be digitized with the help of Blockchain. Blockchain being a decentralized and transparent system will remove any chances of fraud or disputes relating to ownership of land or property. Any changes on the Blockchain have to be made with the consensus of all nodes and therefore rule out the possibility of unlawful encroachment of land. Property disputes are a big problem, especially in India. A very high number of pending cases in courts are those pertaining to property disputes. Blockchain can make the entire system much smoother and more efficient. Records on the Blockchain can clearly prove ownership. Business transactions relating to the sale of land or property are complicated matters as they require hiring a lawyer to certify the deal and various permissions from the government. Registering a sale of property requires the parties involved getting various documents approved and filing records from more than one government department. In the case of Blockchain, the entire process can be handled by pre-coded smart contracts which can handle all the aspects of the sale from taxes to changing the ownership registration of the property involved and make the process very simple for parties involved while minimizing the cost for the government as they do not have to hire as many people.

This will also reduce property tax evasion as the entire information will be available at single sources and smart contracts can take care of taxes themselves. The entire process of paying property tax or stamp duty in case of sale of a property can be automated and funds can be transferred automatically from the user's wallet to the concerned government body. In the case of a property sale, the details of the new owner can be automatically updated and there is no need for the seller or new owner to get the details changes at revenue departments manually.

## 8.9 Education

Academic records, workforce records, instructive endorsements, and so forth, are key resources in the schooling area. Owners or holders of such documents might need to share or prove their ownership to third parties such as educational institutes where they are seeking admission or to organizations where they are applying for a job. Such documents like degrees, diplomas, or academic recommendations can be stored with the utilization of Blockchain innovation. Blockchain can likewise improve on authentication confirmation and check. Provisions for an owner to grant temporary access to such documents for others can be made. Since they are present on a trusted platform, such documents can be easily verified and authenticated. Forging false certifications would also be impossible as only permissioned nodes can issue new certificates.

Creating a distributed repository containing academic records and records of academic qualification can provide an easy and fraud-proof way for employers to review the qualifications of people they are about to hire. This solution also gives the added benefit of secure storage for academic documents like college degrees or mark sheets, as Blockchain offers a very persistent data storage. Students can forget about the hassle of keeping the physical copies of these documents safely as a digital version on the Blockchain will always be present. Blockchain will also eliminate the possibility of forgery of degrees or other documents pertaining to education as well as tampering of mark sheets.

## 8.10 Healthcare

Blockchain innovation has been acquiring perceivability attributable to its capacity to improve the security, unwavering quality, and strength of distributed frameworks. A few regions have profited by research dependent on this innovation, like account, distant detecting, information examination, and medical services. Information's unchanging nature, security, straightforwardness, decentralization, and distributed records are the fundamental highlights that make Blockchain an alluring innovation. In any case, medical services records that contain private patient information make this framework extremely muddled in light of the fact that there is a danger of a security penetration.

Health records of patients can be securely stored on a Blockchain and in case of an emergency, the attending medical professionals can check the entire medical history of the patient just by accessing the patient's account. In case someone decides to visit a new doctor or is referred to a different facility, their entire records can be simply transferred at just a few clicks.

Apart from patient data, Blockchain can also be used to track and manage the availability and logistics of drugs. Procurement of drugs can be automated when a smart contract detects that the stock has gone below a certain level. Blockchain will also increase accountability of hospitals in matters of management of drugs and their use.

## 8.11 Parking Management

With the advances occurring in the space of urbanization, with the increase in urban population density and an increased number of vehicles on street, discovering a parking spot has become one of the significant problem areas for the residents. This is essentially because of the limited number of parking spaces and trouble in tracking down an empty parking spot during occupied hours. Blockchains can be used along with sensors and IoT to track vacant parking spots and also automate the payment process where cameras or infrared (IR) sensors can be used to determine when a car has been parked at a certain spot and automatically trigger a smart contract execution to charge the parking fee. This likewise has the additional benefit of creating income from their generally unused property. Smart contracts can make the entire process simpler for all parties involved and reduce the cost as the number of workers required to run a parking facility can be reduced.

## 8.12 Resource Management

Development and urbanization have put a lot of pressure on our limited resources. There is a need for a feasible and smart way for dealing with the resources intelligently. Resources are essential for the functioning and economy of any city. Increasing population means that we must make better use of the limited resources available. The Blockchain can give a trusted resource monitoring and management platform and furthermore secures any related sensitive information. Using Blockchain can help the framework used in the management of various resources like water quality checking and tracking consumption. Essentially, the bodies responsible for the management of different resources can likewise profit from Blockchain.

Another benefit of using Blockchain is the transparency in the supply chain of essential resources and commodities.

## 8.13 Resource Management

Agriculture is an important aspect of the sustainability of life, as our food production comes from agriculture. Nonetheless, over the past few years, the agricultural sector is facing various issues like climate conditions, soil erosion, monetary issues, and lack of modernization. By including Blockchain in managing the agricultural supply chain and managing other aspects of agriculture, we can somewhat solve these issues. Using IoT devices along with Blockchain, important agricultural data can be recorded and shared with other concerned parties. For instance, a farmer might need monetary help for the harvest. By using a Blockchain and IoT-powered agricultural tracking system there is no necessity for a field official to go to the field to check whether the farmer is giving the correct data. This will be simple for both the farmer and field officials to complete work rapidly. Blockchain can also help to simplify the process of crop acquisition by the government. The supply chain and crop production data can be made open and transparent for everyone so that all parties involved can benefit from it.

## 8.14 Limitation of Blockchain

Blockchain is a promising technology but it is not without limitations. The technology is still in its infancy and various improvements are needed to make it faster and more reliable. Some of the limitations and drawbacks of Blockchains are:

- **Scalability:** Blockchains are not scalable as their traditional counterparts. Running a Blockchain can require a very high number of computational resources, which somewhat limits its scalability. Another issue is that it can be difficult to attract new miners and nodes to a Blockchain network as the cost of operating one can be a hurdle for a lot of people. As the number of transactions increases, these networks will in general turn out to be more slow, costly, and unreasonable for utilization in cases that require speed. That is the reason scaling is turning into a problem in the wider adoption of Blockchain frameworks.

- **Storage requirements:** This is the biggest problem faced by Blockchain technology and limits its scalability. Data on Blockchain has to store on multiple nodes and thus leads to data redundancy. Storage many times the size of the data itself is required to store it on the network. Therefore, storage on the Blockchain comes at a premium and the gas price can be very high. This also makes transactions on the Blockchain slow and expensive. Recent trends like off-chain storage and the use of permissioned networks have reduced this problem to some extent and it might not pose such a big hurdle in the future.

- **Requirement of specialized hardware for mining:** As Blockchains get more and more new miners, the difficulty of mining has to be increased accordingly in order to keep the rate at which blocks are added constant and stop spamming of the Blockchain. Currently, most Blockchains use a proof-of-work algorithm for arriving at a consensus. Proof-of-work requires a lot of mathematical calculations which can prove too complex for a normal processor. In the case of bitcoin, miners were able to profitably mine with normal hardware in the beginning when mining difficulty was low. Later on, they had to depend on more advanced graphic processors which are better suited for this kind of calculation. Nowadays, it is almost impossible to mine with regular hardware on popular Blockchains like bitcoin or Ethereum. Much specialized hardware designed specifically for this purpose is required. Specialized hardware tends to be very expensive and has very high energy consumption. This makes mining on Blockchain inaccessible to many. Critics of Blockchain often point out that all the processing power required for running Blockchains could be put to better use like research in medicine or other sciences. They are also quick to point out the effects on the environment due to high energy consumption. There is a need to come up with new algorithms for consensus on Blockchain, which is not so calculation intensive. Currently, proof-of-work is the only reliable way for all the nodes to arrive at a consensus. Other algorithms like proof-of-stake have been proposed to solve the flaws of the proof-of-work algorithm but as of now, it is incomplete. It is expected that proof-of-stake can replace proof-of-work very soon as the main consensus algorithm and researchers all over the world are working on it. Major Blockchains like Ethereum have already proposed to replace proof-of-work with proof-of-stake in the next version of its protocols.

- **Lack of domain experts:** Blockchain is a relatively new technology and as of now, it has not found a strong footing in sectors other than finance and cryptocurrencies. When compared to other fields of technology, the experts and practitioners of Blockchain are relatively fewer simply because it has not been out there for long enough to find mainstream usage. There are very few college or university-level courses that teach about Blockchain. However, this is fast changing. As cryptocurrencies are becoming more and more popular, people are paying more attention to other applications of Blockchain and realizing the potential of the technology. Technology and IT firms are adopting Blockchain more and more and are training their employees in the field of Blockchain. This is leading to a sharp rate of increase of Blockchain developers and practitioners. Lack of domain experts might seem like a problem for now but in a few years to come, this will no longer be a problem. The new technology takes some time to gather enough practitioners to make it go mainstream. This is also the case with Blockchain.

- **Resistance from governments around the world:** Blockchains like bitcoin and ethereum networks are completely free from government regulations of any kind. The Blockchain is governed only by the principles which have been agreed upon by its members. This makes tracking real identities of people making transactions on Blockchains very difficult for government agencies. The inability to control and regulate them makes governments across the world apprehensive of the technology. Cryptocurrencies have faced opposition from regulators and governments in the past and continue to do so at present. The Indian government banned its citizens from using crypto currencies in the past. However, that ban was lifted shortly after. Recently, China banned all cryptocurrencies for its citizens. Governments fail to realize that there has to be an incentive for miners to join a network and therefore an underlying currency is essential for any public Blockchain network. To access the great technology that these Blockchains provide, we have to embrace cryptocurrencies as they are the driving force behind any public Blockchain. Not all governments have been apprehensive of cryptocurrencies and Blockchain. Russia, for example, has been actively involved in promoting Blockchain as far as moving some services in Moscow completely to Blockchain. The creator of Ethereum, Vitalik Buterin, happens to be a Russian citizen. Such constructive efforts from governments can help in nurturing Blockchain technology and fast-tracking its applications in the real world.

- **Network congestion:** When the traffic on a Blockchain grows, the network can get congested as proof-of-work requires a lot of calculations to mine a new block. The queue of requests keeps getting larger when a lot of people join a Blockchain but the rate of mining has to be kept constant by adjusting the difficulty of mining to prevent drop in the value of the underlying currency and also to keep up with the advancements made in the field of mining hardwares. Requests for transactions by users can take a lot of time before they are included in a block which is accepted as the maximum number of transactions per block has a limit on each Blockchain. This makes utilization of Blockchain for real-time purposes difficult and slow. Whenever a Blockchain becomes immensely popular, it is bound to face this problem. This occurred with bitcoins and Ethereum, the two most popular Blockchains when they were widely accepted. A similar situation arose when crypto kitties, a card collection game on the Ethereum Blockchain, became a hit. Therefore, we can agree that the popularity of a Blockchain can become a bane in disguise which has serious consequences on its network traffic. Various solutions for this problem have been suggested. Some of them are increasing the maximum number of transactions per block, finding ways to use minimum bits to store data and using alternate consensus algorithms like proof-of-stake.

## 8.15 Conclusion

Although Blockchain faces many limitations as of today, innovative solutions like permissioned networks and algorithms like proof-of-stake can make it more scalable and energy-efficient in the future. When paired with IoT in Smart Cities, Blockchain can take them a step further into creating an ideal society. Smart cities are utilizing Blockchain as a means to upgrade many aspects of metropolitan living. Many cities have already started using Blockchain for various purposes. Recently, Moscow conducted an election on a Blockchain-based e-voting system.

Blockchain is a promising innovation. Blockchain, especially when utilized working together with different advancements, offers associations a chance to reexamine their inner and outside measures, eliminate shortcomings, improve straightforwardness and provenance, and construct a superior association generally. Nonetheless, it faces various difficulties that could influence its reception across associations. Be that as it may, altogether more examination is needed to defeat the extensive administration and mechanical difficulties included in that.

In the present day, there is no solution better than Blockchain that exemplifies trust. It is an expectation and assumption that this article will help keen city organizers, engineers, draftsmen and scholars execute Blockchain as the epitome of trust in brilliant urban areas that are progressively getting advanced.

Smart city pioneers are embracing Blockchain innovation as a cutting-edge technology that can connect different services to share information, reduce security breaks, give consistent exchanges, and offer transparency for public administrations. Simultaneously, Blockchain likewise offers energizing freedoms for a smart city's private businesses. In this paper, the current limitations of Blockchain have been discussed in brief but these limitations should not discourage cities from adopting Blockchain as these limitations are either very little or solvable in the near future.

These problems might seem like a great roadblock in the future of Blockchain technology. For the potential that Blockchain offers, it can be safely assumed that it is well worth investing in finding ways in which these problems can be overcome. Tech companies all over the world are investing heavily in this technology as they can see the potential it offers. Researchers all over the world are working to make Blockchain more efficient in terms of energy consumption as well as storage optimization. As Blockchain finds more and more mainstream applications and adoption, new solutions and improvements will keep coming more frequently. The potential offered by Blockchain when paired with various systems present in Smart Cities is virtually limitless. From waste management to healthcare and logistics management, there is almost no field which can be improved upon when it comes to Smart Cities. Sharing arrangements and information among regions inside Smart Cities by utilizing Blockchain innovation will drastically develop and considerably improve the way of life all throughout the planet. The IoT can upgrade our metropolitan way of life, accommodating the execution of distributed processing for constant information analysis and quick activity and response. It can be safe to assume that in the near future, Blockchain will play a much larger role in the Smart Cities.

## References

1. Yaga, D., Mell, P., Roby, N., & Scarfone, K. (2019). Blockchain technology overview. arXiv preprint arXiv:1906.11078.
2. Davidson, S., De Filippi, P., & Potts, J. (2016). Economics of blockchain. Available at SSRN 2744751.
3. Tasatanattakool, P., & Techapanupreeda, C. (2018, January). Blockchain: Challenges and applications. In *2018 International Conference on Information Networking (ICOIN)* (pp. 473–475). IEEE.
4. Chen, S. Y., Song, S. F., Li, L. X., & Shen, J. (2009). Survey on smart grid technology. *Power System Technology*, 33(8), 1–7.
5. Christidis, K., & Devetsikiotis, M. (2016). Blockchains and smart contracts for the internet of things. *IEEE Access*, 4, 2292–2303.
6. Buterin, V. (2014). A next-generation smart contract and decentralized application platform. *White Paper*, 3(37), 2588.
7. Dorri, A., Kanhere, S. S., Jurdak, R., & Gauravaram, P. (2017, March). Blockchain for IoT security and privacy: The case study of a smart home. In *2017 IEEE international conference on pervasive computing and communications workshops (PerCom workshops)* (pp. 618–623). IEEE.

8. Kshetri, N. (2017). Can blockchain strengthen the internet of things?. *IT Professional*, 19(4), 68–72.

9. Batty, M., Axhausen, K. W., Giannotti, F., Pozdnoukhov, A., Bazzani, A., Wachowicz, M., ... & Portugali, Y. (2012). Smart cities of the future. *The European Physical Journal Special Topics*, 214(1), 481–518.

10. Albino, V., Berardi, U., & Dangelico, R. M. (2015). Smart cities: Definitions, dimensions, performance, and initiatives. *Journal of Urban Technology*, 22(1), 3–21.

11. Zanella, A., Bui, N., Castellani, A., Vangelista, L., & Zorzi, M. (2014). Internet of things for smart cities. *IEEE Internet of Things Journal*, 1(1), 22–32.

12. Liu, Y., He, D., Obaidat, M. S., Kumar, N., Khan, M. K., & Choo, K. K. R. (2020). Blockchain-based identity management systems: A review. *Journal of Network and Computer Applications*, 102731.

13. Toyoda, K., Mathiopoulos, P. T., Sasase, I., & Ohtsuki, T. (2017). A novel blockchain-based product ownership management system (POMS) for anti-counterfeits in the post supply chain. *IEEE Access*, 5, 17465–17477.

14. Niranjanamurthy, M., Nithya, B. N., & Jagannatha, S. (2019). Analysis of Blockchain technology: Pros, cons and SWOT. *Cluster Computing*, 22(6), 14743–14757.

15. Rubtcova, M., & Pavenkov, O. (2018, February). Disadvantages of Using of Blockchain in Electronic Election in Russia. In *Blockchain & Bitcoin Conference*, Bengaluru, India (Vol. 22).

# 9

## *Mutual Surf-Scaling Factor Features Analysis for Detecting DDoS Attacks Based on Lattice Behavioral Using Spread Spectral Soft-Max-Recurrent Neural Networks*

### N. Umamaheshwari and R. Renuga Devi

*Vels Institute of Science and Technology Advanced Studies (VISTAS) Vels University, Chennai, India*

## CONTENTS

## 9.1 Introduction

Distributed network attacks are services that attain attacks during network data transmission and this becomes more vulnerable in cloud computing. Most DDoS service attacks are handled with multiple requests and send multiple requests beyond the network resource site designed to attack from running the site to prevent intrusion from attackers properly. Analyzing the service attacks in cloud service representation requires stringent feature analysis because attackers create more non-related service representation by increasing more features and dimensionality. Owing to this complex nature, each feature is analyzed to test the relative margins. So, testing is an important step against DoS/DDoS attacks.

DOI: 10.1201/9781003206088-9

However, the high variability of the DoS/DDoS attack type is the problem in detecting such attacks. Most case problems are non-related feature weights prediction in transmission point of packet processing rate.

Distributed services are typically affected by intruders and they use a variety of computer resources to launch an attack against one or more target coordinates during the data transmission. Denial of service (DoS) attacks may deny access to shared services and resources to other legitimate users. Their main purpose is to extract relative features using surf-scale feature selection and apply the deep neural network (DNN), based on retrieval of features from the neural network unit. When training large datasets, Soft-Max activation depends on recurrent neural network (RNN) which has been proven to be in feature analysis. In particular, by learning the sequence of attacks until deep long-term and short-term DDoS attack methods are detected, the benefits of network packets and large network traffic are valued.

Refused Service (DoS attacks) occur mostly by non-verification service resemble at transmission medium which easily affects any client system. Many attacks are based on malicious purposes, such as the destruction of certain corporate services in exchange for money by performing the targets. DoS or DDoS attack occurs, because DoS attacks magnified form, known as zombies, misleads facts on the Internet directly to thousands of susceptible hosts attacker. There are many types. The use of legitimate third-party components hides the identity of the attacker ruins of DDoS attacks. In the reflection-based DDoS attack, the attacker sets a desired target internet protocol (IP) source, and IP address of the victim as the response to the packet forwarding packets overwhelm the victim reflector server. DDoS attacks and challenges are based on the reflection of complicated data communication. An intrusion detection system (IDS) detects all aspects of networks and systems from malicious activity or policy safety nets.

IDS things are hybrid devices and due to the increasing popularity of the internet, the safety experts are concerned about protecting a combination of two or more IDS approaches. Therefore, proposing an alternative method of detection based on long short-term memory (LSTM) can take advantage of deep learning in a successful DDoS attack detection of new types of malicious packets. A classification is based on the behavior detected by the traffic potential of DoS attacks, combining the advantages of the hybrid IDS method to overcome the defects of the IDS signatures and anomalies.

Deep neural network-based DDoS attack detection is implemented for identifying the feature learning, defense in depth. This improves performance by identifying DDoS attack traffic. The development of a classification sequence DDoS attack detection and packet-based DDoS detection conversion is based on the detection window. Automatically, feature analyses are carried out using DNN from high-level, low-level ones, powerful expression, and reasoning. This has facilitated the designing of a learning mode relapse DNN network traffic from network attacks and a tracking array. Experimental results show superior performance compared to traditional machine learning models of the model.

DDoS attacks dynamic time may indicate that using alternative Deep Learning (DL) architecture combined with Convolutional Neural Network (CNN) suggests a better LSTM novel optimization method. The model can maintain a context for each time the overall flow to push data over a network. Make full use of the cellular neural networks by learning DDoS attacks and benign traffic behavior which apply to limited online resources. Both are based on the low processing overhead of DDoS detection system during runtime environment and attack detection.

## 9.2 Related Work

DDoS attacks are not natural as they attack transmission packets under false time. malicious activities [1] and distributed denial of service (DDoS) detection attacks are an effort to mark a non-feature observation that cannot use their target client to make well-known security attacks.

The Software Defined Network (SDN) is of great interest as the new paradigm of the network. As can be seen, SDN security is very important [2]. The Distributed Denial of Service (DDoS) detection attacks have become a plague on the Internet.SDN as a central point, centralized data management and traffic management [3]. SDN and open-source products are vulnerable to security threats. It is also a security policy when the controller is forced to attack the most.

The SDN provides the control, and the data plan is separated. Then, the controller has a central controller in the whole communication system. SDN provides the ability of the network programming which can dynamically create policies flow [4].

The network (SDN) definition of agile software has become an important promoter of the future architecture of the Internet [5]. Despite this function commitment, SDN is inherently insecure. Defined network (SDN) software is a promising network model, which provides great manageability, scalability, controllability, and flexibility [6]. However, the flexibility is provided by SDN architecture by revealing some new design problems in network security.

Intrusion Detection plays a significant role in confirming data security key expertise which precisely recognizes more attacks in the network. The RNNs have been more successful in image processing [7], especially in the advanced vision applications of natural language processing, such as identifying and understanding. However, they add very little value to solve the problem of information security, such as the detection of such attacks.

A DDoS detection attack uses the client/server technology of multiple computers in one or more target combinations of launching attacks to increase the strength of the attack [8]. Almost every year, this attack holds the highest score of cost of the economy as a whole than all other attacks. Machine learning (ML) is based on the IDS to cure global economic growth and reduce the incidence of network incidents, such as DDoS attacks.

The conventional IDS methods have some limitations including the supervised learning method that requires a relatively low false alarm detection rate. The number of data-tags-based unsupervised learning algorithms used to solve these problems [9] exacerbated the existing network attacks, including services like ICMPv6-based denial of service (DoS) attacks and service (DDoS) attacks and distributed denial variants [10].

Service-attack DoS attacks are one of the most common attacks. DoS attacks are difficult to alleviate at low rates [11] and are becoming more difficult to distribute denial. Low-Rate Denial of Service (LDOS) uses loopholes in the Transmission Control Protocol (TCP) congestion control mechanism through the low, fixed interest rates which affect the sending of malicious traffic to the victim's machine [12].

One of the most ruthless attacks is an important dispersion of DoS attacks. This kind of attack tools are used every day as technology improved [13]. Extensive research has been carried out on denial of service (denial of service) and DDoS attacks, relief, attacking, and it has been trying to mitigate such attacks. For example, a low-rate DDoS (LR-DDoS) attack is known as indistinguishable [14].

It cannot be used to attack the outgoing statistics. It is only the passive defense after the attack, and these methods are difficult to go back and track down the attackers [15]. However, from malicious attacks, it is ensured that the system of things can be a daunting task. One of the most common malicious attacks is a denial of service (DoS) [16], distributed denial of service (DDoS) attack.

A denial of service (DoS) attack is an attack on computers in today's online world that is now very common [17]. Automatically, they detect the type, such as the DDoS attack packets, before dropping through and prevention is the best [18,19]. Cloud computing provides greater flexibility, less maintenance, end-user demand infrastructure costs, virtual resources, and scalability than the IT revolution in technology [18].

When detecting DDoS attacks, a general deterioration of the detection performance is due to the small sample size [20]. The detection of this small sample of DDoS attacks s based on a deep transfer of learning. These attack methods, mostly IP and TCP layer DDoS attacks, are not suitable to detect application-layer DDoS attacks based on the request of flood attacks [21]. However, application-layer DDoS attacks include requirements of flood session and asymmetric attacks.

A DDoS attack is one of the most expensive. This article discusses machine learning algorithms to detect DDoS attacks use [22]. Traditional support vector machine learning algorithms and new depth, deep feed forward (DFF) are valuable. They will automatically drop packets before the attack by DDoS attack and can detect the type which is the best prevention [23]. Traditional solutions only monitor and provide feedback based on feed-forward solutions instead of machine learning.

## 9.3 Mutual Surf-Scaling Factor Features Analysis for Detecting DDoS Attacks

The proposed method's objective is to detect DDoS attacks using mutual surf-scaling factor features analysis ($MS^2F^2A$) based on lattice behavioral using soft-max spread spectral recurrent neural network (Sx-$S^2$RNN). Initially, the collective data set is preprocessed to reduce the dimension of the dataset, which is the non-redundant data. Then, the mutual surf-scaling features are applied to filter the relational features based on lattice behavioral access rate (LBAW). Further, mutual scaling features are observed by distance vector space to choose the marginal cluster as features group. Then, an inter-segment soft-max activation function is proposed to create a logical neural network to predict and optimize the output class. The feature selection supports for relation compound extraction to reduce the non-dimensional nature of DDoS data. Further, the classifier needs to classify the result to construct spread spectral recurrent neural network ($S^2$RNN) to predict the DDoS labels based on the relative spectral values used to detect the DDoS attack detection.

The spectral values are marginalized as influence weight that is trained with activation logical neuron satisfaction in the input signal. Attention-Based Recurrent Neural Network (ARNN) retains the redundant features of unattended feature learning techniques. Figure 9.1 shows the Soft-Max Spread Spectral Recurrent Neural Network (Sx-$S^2$RNN). It has taken the redundant features for attaining the best classification. After preprocessing, the actual output is compared to the input if required. Then, the error is sent back to the compatible system. During training, the data is processed several times, so that the network can adjust the weights to predict the classification results.

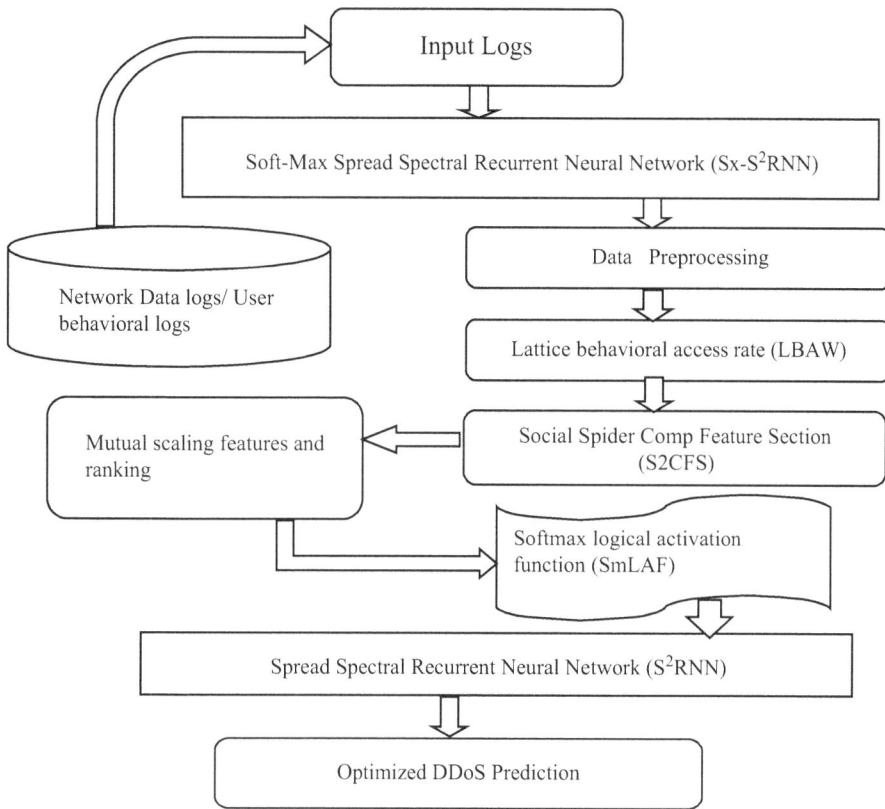

**FIGURE 9.1**
Soft-Max Spread Spectral Recurrent Neural Network (Sx-S²RNN).

### 9.3.1 Data Preprocessing

In this stage, the presence of collective dataset based on network behavior condition observes DDoS dataset. This verifies the present feature cases, labels counts, and missing values prediction. This can easily treat it as an outlier training model and convert it to a normalized negative value. The second step is to create class labels before using data redundancy to reduce the noise from the raw dataset. Noise-less data during the preprocessing phase is to be created to create the data labels for the features point.

---

**ALGORITHM 9.1  Preprocessing dataset**

1. Start
2. Initialize the Collective DDoS datasetDrf and user behavioral category dataset Uwd
3. Input dataset Drf → cr1, cr2, cr3 + Uwd (Index +1)
4. For (Drs combines attributes as feature Fi)
5. Check → Null field if Yes
6. Terminate record and update recordsetRcs.
7. Else

8. Return
9. Retire Rts ← Crs+Wd;
10. End if; End for;
11. Compute Duplicate and non-fill case Fls
12. Remove duplicate if Rd → Rcs
13. End if
14. Check if (Rcs!=Empty fill near average)
15. Get All the future index count terms.
16. Returns = $F(Crf + Wd)^n = \sum_{I=0}^{n} \binom{n}{k} Rcs.Update$ → get real index, where n is total

dominant features with k observed feature.
17. Return redundant feature set-Rds
18. End if
19. Stop

The above Algorithm 9.1 prepares to process the original data for noise reduction, and the collective behaviors are collected as features that are ordering the feature index with redundant features. There are just two strategies: accurate and approaching for detecting DDoS attack with dynamics that need to be processed by removing the non-redundant terms.

### 9.3.2 Lattice Behavioral Access Rate (LBAW)

In this stage, the network user behavioral features are observed through access control on serving optimality. This estimates the average mean weight of relational term present in DDoS features. Consider the compound relation features $A = \{a_1, a_2, \cdots, a_n\}$ and $B = \{b_1, b_2, \cdots, b_m\}$ and the entropy is calculated regarding the user's behavioral activities to be calculated by input relational $Xi$ and $Xj$ features relation, then Rds → $\Sigma i = 1np(xi) = 1\Sigma i = 1np(xi) = 1$ and $\Sigma j = 1mp(Xj) = 1\Sigma j = 1mp(Xj) = 1$ by the heuristic feature (Hf) is the representation given below:

$$\text{Hf}(A) = \sum_i^n \quad \begin{matrix} ni(Xi) = 1 \text{which is} \\ \text{equeivalent another input series } P(Ai)\log 2p(A2) \end{matrix} \tag{9.1}$$

and

$$\text{Hf}(B) = \sum_j^m \quad \begin{matrix} mi(Xj) == 1 \text{which is} \\ \text{equeivalent realtional input series } P(Bi)\log 2p(B2) \end{matrix}. \tag{9.2}$$

The joint features from sources A and B state are equivalent, $\{A, B\} = \{a_1b_1, a_1b_2, \cdots, a_nb_m\}$, this creates joint distribution beyond the absolute weightage between the two states $\{A, B\}$ $\begin{vmatrix} a_1b_1 & a_1b_2 & \dots & a_nb_m \\ p(a_1b_1) & p(a_1b_2) & \dots & p(a_nb_m) \end{vmatrix}$. The joint feature data correlation among the behaviors is the entropy value of A and B:

$$\mathrm{Hf}\left(\mathrm{A,B}\right) = 1 - \sum_{i=1}^{n}\sum_{j=1}^{m} p\left(a_i b_j\right)\log_2\left(a_2 b_j\right) \tag{9.3}$$

The independent features from A and B are the Relation Feature (Rl), defined as

$$\mathrm{Rl}\left(\mathrm{A,B}\right) = \mathrm{H}\left(\mathrm{A}\right) + \mathrm{H}\left(\mathrm{B}\right) - \mathrm{H}\left(Ab\right)\text{real probability}\,(0,1)\,\text{joint relation} \tag{9.4}$$

The entropy value is the correlation between optimal features referred to as relevance features 0 and 1. The correlation relates the A and B independent relational features from sources A and B that welted the dependency of joint relation. These joint relation attributes are considered redundant features for further classification.

### 9.3.3 Social Spider Comp Feature Section (S2CFS)

The further objective is to improve the feature selection based on mutual surf-scale nested feature clustering. Accuracy for considering related features is dented with interest analyses. To improve the clustering performance, efficient nested integrated features are presented. The mutual social spider comp feature section (S²CFS) technique evades calculating the distance of each data feature occurrence to the cluster centers again and again and saves the running time.

---

**ALGORITHM 9.2    Social spider comp feature section (S2CFS)**

Input: Featured attributes Rds → R(*T*) as H(*a, b*)
Output: Decision rule out feature selected dataset values
Step 1:  Start
Step 2:  Initialize the Kdd-cup from LBAW term R(*T*)
Step 3:  Compute the tags as margins scaling threshold values K features
Step 4:  Evaluate the feature attain weight in attributes Term setTs, K.
  //compute the K-number of features as iterative levels for selection
  Estimate fitness For Fixed member as Performance Selection (PS) relevance weight
Step 5:  Fix the scaling value to search behind factor Max-Min
  Compute relevance measure of Relative margins weight

$$\mathrm{Ps} = \left(Ts\right)^n = \sum_{k=0}^{n} \mathrm{max-min}\binom{n}{k}\text{Initilazed feature}^k K$$

  Observing Number of features related to Margin scale on traffic
  Obtain K-number of relative margin-surfed features
Step 6:  Compute the feature relevance generated term (Tg) features—Tg
  Retain the extracted relevance term Et ← Tg
  Estimate the real featured Limits processed term Pt, Non-real Nt
  Check if (Tg as Et →)
  Select to surf-scale marginal transition feature Tg
Step 7:  Return Tg
  End if
  End

\

Surf-scale nested clustering structure predicts data in every iteration, which is to be used in the next iteration for semantic relation. This relevantly measures the relative features closeness of the cluster group. The average weights measures are categorized into class variables and domain values as data points. The data points with high confidence and low support in the weightage will be ignored. Based on the measure estimated, the method selects a single cluster to which the data point must be indexed.

### 9.3.4 Mutual Scaling Features and Ranking

These DDoS value variables have identity links to clear the relational connectivity between the DDoS value features. This reduces the non-relational terms and the cleansing way and combines distortional features same as relational attribute variables, to reduce the non-relational terms.

---

**ALGORITHM 9.3    Mutual scaling features and ranking**

Input: feature process set S²CFS as Ps ← Prds (Tg)
Output: Relative weightage Rds set
Start

Step 1:   Initialize the processed set Ps
              For each feature (read ← DDoS value id)
Step 2:   For each attribute Ca from Ps
                  Identify the count terms Ct value
                      Max term value (max variable count → confidence)
                      Rearrange the DDoS value Id
                      Create links between term DDoS value id → cd1,cd2,..
Step 3:   compute the key terms sentence kt
                  For each DDoS value feature kt from Ps
                      For each DDoS value term Kt from Ps
                          If Kt € max term then
                              Identify max count relation with other attributes.
                              Relation Set Rs = ∑(Concepts€Kt)+Ps.
                          Compute the Number of attribute relations it has.
Kts = ∑ Relations €Gi similar term
                              Compute the max count attribute value
mval = ∑ feature Links(Kt) < –∑mval(Kt)
                                      Create link Kt identified relative feature links
                                      Relative link Rt ← kt+ca;
                          End
                      End
              End
Step 4:   compute the ranking conceptual feature links → CRL.
              For each feature ← RL
                  CRL = = ∑Concept (Links (Kt))) € ∑Concept (Ca)!=Ps
                  Compute relative feature → RLK
                      RLK = (Kt + Ca) + NIL

> Add to feature link set CRL ← RLK;
> End
> End
> Stop.

The above Algorithm 9.3 reduces the frequent intervals between the connected attribute links that are related to each other. The maximal DDoS value has count which refers to the repeated DDoS values. So, distinct frequency will be evaluated to originate the data.

### 9.3.5 Soft-Max Logical Activation Function (SmLAF)

At this stage, the classifier finalizes DDoS attacks' result by the risk of Intrusion and its categorized class based on the sigmoid activation function. The biased weight control initialized all the feature weights feed into the hidden layer through the Soft-Max activation function. The activation function trains the neurons to tune the resultant clusters into mean depth values to create a logical rule for the neurons.

The feed-forward network is optimized with an adaptive forager search algorithm rule obtained as a training rule into a recurrent neural network for tuning neurons to the closest weight prediction. This constructs a hidden layer with logical output function evaluation for adjusting weight depending on the marginal class from each iterative feature analysis up to optimal prediction with sinusoidal representation. Based on the sinusoidal weight, the maximum features weight is attained to the neural decision at a gradually increased manner if the redundant weight finds that they spread the value into the nearest class without backpropagation to avoid link breakage during training. This reduced search optimization of relevance feature feeds forward into the next layer.

The activation function trains remain as $f(x) = \begin{cases} y = 1 \text{ if } \sum_{i-1}^{n} w_i x_i \geq b \\ y = 0 \quad \text{otherwise} \end{cases}$, where $f(x)$ remains the logistic activation of neuron trained with intraclass logistic transformation, $w_{(t+1)} = w_t - \mathbb{N} \Delta w_t$ and $b_{(t+1)} = b_t - \mathbb{N} \Delta b_t$. The training function is Iterated till the neuron gets the closer mean value

We begin to function W approximately as b. Tracking all the information using each corrugation function to find and repeat the corresponding prediction result, find, and calculate the square error loss.

A fully integrated feed-forward of the neuron is illustrated in the construction of the network

$$net_{i(t)} = \sum_{j=1}^{j} w_{ij} y_{j(t)} + x_{i(t)}, i = 1 \dots j \tag{9.5}$$

and

$$Ti \frac{dy(t)}{dt} = -yi(t) + \varphi(net_i) + x_{i(t)}, i = 1 \dots j. \tag{9.6}$$

The frequent neurons of weightage are optimized in constant τi based on the current unit is a valid training weightage based on the interclass logistic activation function. The input X(i) and Y(i) bet is activated based on the weightage w(i).

---

**ALGORITHM 9.4   Soft-Max logical activation function (SmLAF)**

Input: Its features as Current Sample scs, Adapted feature *rule net*$_{i(t)}$ $- \rightarrow$ Apt.
Output: optimized class pattern

Step 1:  Start: compute the DDos Max rate and occurrence of feature Max weight term

Step 2:  Read Apt. Data values and Scs. Data values.
         For each feeded layer class Pc.

Step 3:  Compute the hidden layer neurons weight to c as set $= \displaystyle\int_{i=1}^{size(Apt)} \Sigma\, Apt(i).class = c$
         Closest pattern Pps = Closest pattern (Cset).
         For each closest pattern on the relative link from Influence rate, each pattern p

Step 4:  By each similarity, features are classified as categories based on the risk of max feature categorized class

$$\text{Subcluster feature selection DPfs} = \frac{\displaystyle\int_{i=1}^{size(p)} \Sigma\, P(i) == Scs(i)}{size(p)} \tag{9.7}$$

End

$$\text{Compute cumulative rate DPFS} = \frac{\displaystyle\sum_{i=1}^{size(pps)} Pfs}{size(pps)} \tag{9.8}$$

End
Optimized Computed DDoS Detection (CPR) = PFs return set maximum values

Step 5:  Stop

---

The input features are trained with the classification structures and then labeled with neural weights, representing the classes. The input factors aides in separating the area learning about relational features to the hidden neurons. The activation function tunes each neuron to optimize the accuracy of the classification by class by reference.

The recurrent neural algorithm consists of artificial neurons activating functions that involve a hidden layer due to feature categorization. For each neuron, the logical contains created decisive statements to predict the feature weight. Therefore, repetitive nerve cells spread close weights along the cluster group to find the optimal class based on the nerve structure. The feed-forward approach in the hidden layer constructs the logical condition closest to the feature weight and each neural layer linked forwarded to point relative weights assigned by each neuron. The feature weight is observed based on the activation function to predict the output.

### 9.3.6 Spread Spectral Recurrent Neural Network (S2RNN)

In this stage, learning features are analyzed by the adaptive inter-segment activated recurrent neural classifier based on its input feature progress. All the neurons trained with activation function are called inter-segment sigmoid neural constriction. Neural searches are transmitted between these neurons through connections. Due to this recurrent connection, the links are connected between relative cluster groups. All the hidden layers are fed to a connective group based on the activation functions. The connections with the neurons will be updated to know the progress and the weights associated with the feature index.

---

**ALGORITHM 9.5   Spread spectral recurrent neural network (S2RNN)**

Input: feature values → Fv
Output: Recommended class by DDoS attack
Step 1:   Compute Hidden layer construction As Fv
Step 2:   Create feature index input progress Fip
Step 3:   Construct feed-forward layer with the inter-segment constructor of neuron logic
   for executing fip as redundant link
     Progress the Link spread on relative cluster group
     Sort all the features to index F as cluster Fci

$$\text{Fip} \rightarrow (y+b)^n = \sum_{i=0}^{n} \binom{b}{y} y^k w^{n-k} + c \sum_{l=1}^{n} \text{€}Its \qquad (9.9)$$

Step 4:   Compute for each scaling intensive lattice weight
   For all scaling j in the cluster, ci do
     For all term indexes, modify the weight x
     Check neural layers

  If layer exceed Forward → new xnew, ci, $jw_{(t+1)} = w_t - \mathbb{N}_\Delta w_t(net_{i(t,)})$    (9.10)

Step 5:   Compute scaling if Irpi is the best ci class, then
     Fix margin winner class YnewFsi = b*Relative weight, Fci
    Feature index spread

$$b_{(t+1)} = b_t - \mathbb{N}\Delta b_t \qquad (9.11)$$

    End for each
   End for
  End if
Step 6:   Creating activation for iteration in the hidden layer
   Construct max layer at each end y → Fsi as forwarding weight
   End for
Step 7:   Return Fsi class by reference as recommending class.

---

The above algorithm precedes the classification which is observed from the input features. Based on sigmoid activation, the neurons get trained by closure measure. The forwarding hidden layers processed the closest fitness weight to categorize the class. This produces the best DDoS attack prediction, as well as a class by reference.

## 9.4 Result and Discussion

The testing parameters are evaluated in a network simulator with proficient network traffic predicted dataset with confused matrix estimation using random data processing in a simulation environment. The projected mutual surf-scaling factor features analysis for detecting DDoS attacks is based on lattice behavioral using spread spectral Soft-Max-recurrent neural network. These weights are relatively analyzed on threshold value from feature attacks variance. The proposed system is compared with existing principles of performance methods like SSO, PSO, BPNN, TPF-IEHO, and SFS-C²ONN. The implementation proves the redundant feature-based DDoS detection with higher classification accuracy. The degree of classes is marginalized on actual threshold values from the attack nature of the network environment. This implementation is tested with tested i3 processer in 4 GB of Ramwith Sniffer centric tool during run time execution in actual state.

The KDD-cup is carried out as a network behavior dataset by considering the traffic factors to process the simulation environment having the collected number of values. Table 9.1 shows the environment variables and their values processed.

A confusion matrix progresses the testing results to evaluate the performance of the proposed system by taking the absolute values from the feature values to classify the results. The classification is calculated by

$$\text{Classifaction Accuracy} = \frac{TN + TP}{\left(TP + FP + FN + TN\right)} \qquad (9.12)$$

Classification defines the sensitivity specificity of frequent measure rate predicted by the precision/recall rate produced by positive value with the absolute result by categories class. Figure 9.2 shows the classification accuracy.

The classification accuracy shows the performance of the proposed implementation Sx-S²RNNin Table 9.2. The Sx-S²RNN produces higher accuracy in DDoS detection rate than other methods. Also, the sensitivity rate is analyzed by the actual rate of true-positive values, which is averaged by mean observation of true-positive and false-negative values.

$$\text{The estimation of sensitivity} = \frac{TP}{TP + FN}, \qquad (9.13)$$

**TABLE 9.1**

Environment Variables and Its Values Processed

| Environment Variables | alues Processed |
|---|---|
| Dataset used | KDD-Cup-DDoS |
| Tool used | NS2-OTCL |
| Routing protocol | AODV, UDP |
| Range | 100*100 m |
| Number of features | ≥30 |
| Number of class | High, medium used |

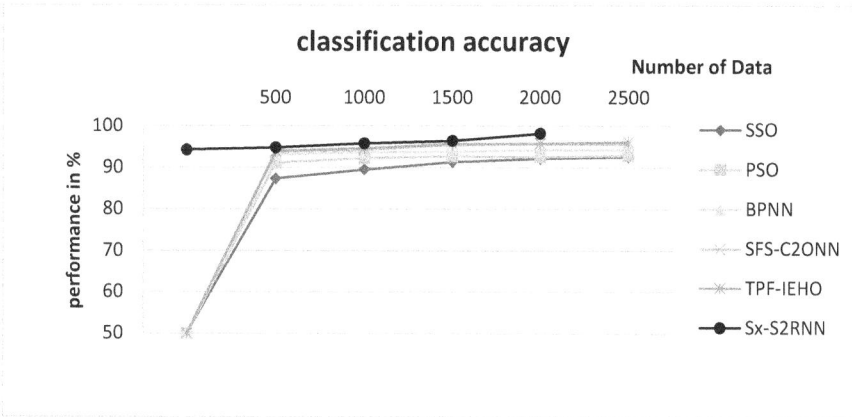

**FIGURE 9.2**
Performance of classification accuracy.

**TABLE 9.2**

Performance of Classification Accuracy

| Number of Data | Performance of Classification Accuracy in % | | | | | |
|---|---|---|---|---|---|---|
| | SSO | PSO | BPNN | SFS-C²ONN | TPF-IEHO | Sx-S²RNN |
| 500 | 87.3 | 91.1 | 93.1 | 93.8 | 93.9 | 94.2 |
| 1000 | 89.5 | 92.2 | 93.6 | 94.2 | 94.6 | 94.8 |
| 1500 | 91.3 | 92.7 | 93.8 | 95.4 | 95.6 | 95.8 |
| 2000 | 92.1 | 92.5 | 94.2 | 95.6 | 95.8 | 96.4 |
| 2500 | 92.6 | 92.9 | 94.2 | 95.7 | 97.9 | 98.2 |

The proposed Sx-S²RNN produces 91.4% sensitivity accuracy compared to the other methods.

Analysis of Sensitivity (Figure 9.3)

The DDoS sensitivity defines the actual rate of feature occurrence in a true-positive rate. This detection estimates the feature depending on true observation weights prediction in the neural classifier. The implemented Sx-S²RNN produces the best performance in feature observation carried out through classification accuracy (Table 9.3).

The proposed Sx-S²RNN system implementation proves the sensitivity analysis in higher performance by prediction feature classification in the DDoS category. Specificity calculates the true-negative values averaged by true-negative and false-positive values.

$$\text{Specificity } \frac{TN}{TN+FP} \qquad (9.14)$$

Figure 9.4 demonstrates the contrast specificity, which is called the precision rate calculated by observing averaged mean of true-negative values. The proposed Sx-S²RNN attains a higher feature classification prediction rate and performs better than other dissimilar methods.

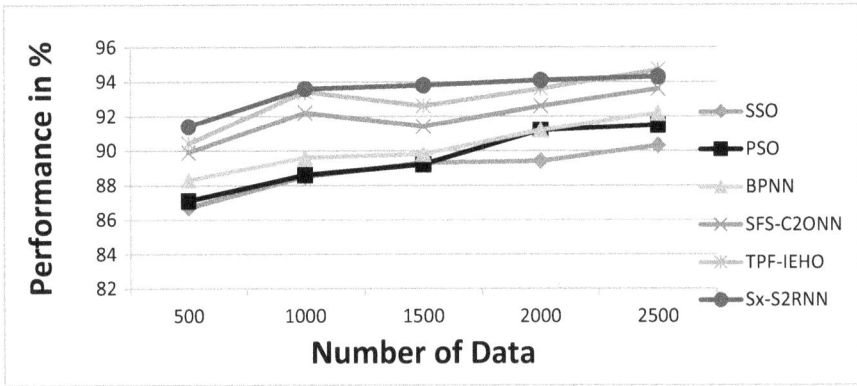

**FIGURE 9.3**
Performance of sensitivity analysis.

**TABLE 9.3**

Performance of Sensitivity Analysis

| Number of Data | Performance of Sensitivity Analysis in % | | | | | |
|---|---|---|---|---|---|---|
| | SSO | PSO | BPNN | SFS-C²ONN | TPF-IEHO | Sx-S²RNN |
| 500 | 86.7 | 87.1 | 88.3 | 89.9 | 90.4 | 91.4 |
| 1000 | 88.5 | 88.6 | 89.6 | 92.2 | 93.4 | 93.6 |
| 1500 | 89.3 | 89.2 | 89.8 | 91.4 | 92.6 | 93.8 |
| 2000 | 89.4 | 91.2 | 91.2 | 92.6 | 93.6 | 94.1 |
| 2500 | 90.3 | 91.5 | 92.2 | 93.6 | 94.7 | 94.3 |

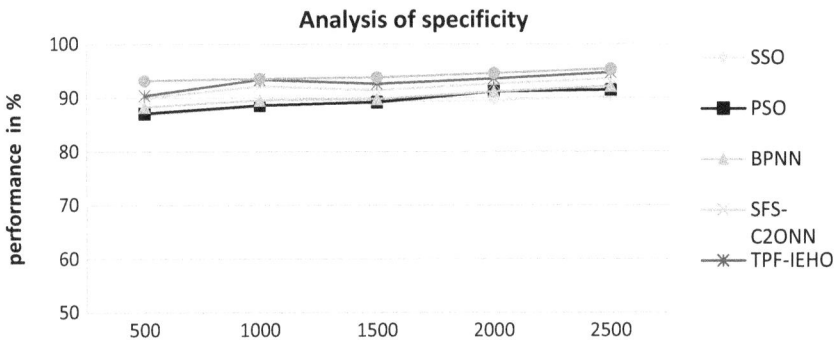

**FIGURE 9.4**
Performance specificity.

The precision measure represents the harmonic representation by avoiding the non-negative features from the true-negative values to average the absolute values depending on the feature weights. Table 9.4 shows various methods of precision analyzed to specify the rate. The proposed method highlights the higher performance compared to other methods (Figure 9.5).

**TABLE 9.4**

Performance Specificity

| Number of Data | Performance of Specificity in % | | | | | |
|---|---|---|---|---|---|---|
| | SSO | PSO | BPNN | SFS-C²ONN | TPF-IEHO | Sx-S²RNN |
| 500 | 82.3 | 87.3 | 89.3 | 91.3 | 92.1 | 93.2 |
| 1000 | 83.8 | 87.6 | 91.2 | 91.8 | 92.4 | 93.6 |
| 1500 | 84.2 | 88.5 | 92.6 | 92.8 | 93.6 | 93.8 |
| 2000 | 85.3 | 88.9 | 92.8 | 93.2 | 94.2 | 94.6 |
| 2500 | 86.3 | 90.2 | 93.5 | 94.5 | 95.2 | 95.4 |

```
Accuracy of the model is:  97.93938291810632
Confusion Matrix:
 [[    82    24]
 [     6 49379]]
Report:
                   precision    recall  f1-score   support

              0       0.93      0.77      0.85       106
              1       1.00      1.00      1.00     49385

       accuracy                           1.00     49491
      macro avg       0.97      0.89      0.92     49491
   weighted avg       1.00      1.00      1.00     49491
```

**FIGURE 9.5**
Accuracy obtained from the confusion matrix.

Based on Specificity and sensitivity, the actual values are estimated by F – measure (False Classification) $= \dfrac{2 * \text{Precision} * \text{Recall}}{(\text{Precision} + \text{Recall})}$. The error rate under 2.5 mean average rate is defined as 97.93% classi fication accuracy. The classification as a non-residual form of original data is classified as wrongly defined as a definite class. The representation of this formula is identified as,
by

$$\text{False Extraction Ratio}\left(\text{Fer}\right) = \sum_{k=0}^{k=n} \times \frac{\text{Total Dataset Failed to Classify}\left(\text{Fer}\right)}{\text{Totalno of Data}\left(\text{Fr}\right)} \quad (9.15)$$

Figure 9.6 demonstrates the contrast of the false sorting ratio formed by dissimilar approaches and the projected Sx-S²RNN technique that has shaped less false classification than the additional remaining approaches.

The above Table 9.5 shows the performance of the proposed system that can prove misclassification by identifying false rates in classification accuracy. These false rates in lower redundancy in proposed Sx-S²RNN produce the best performance compared to the other

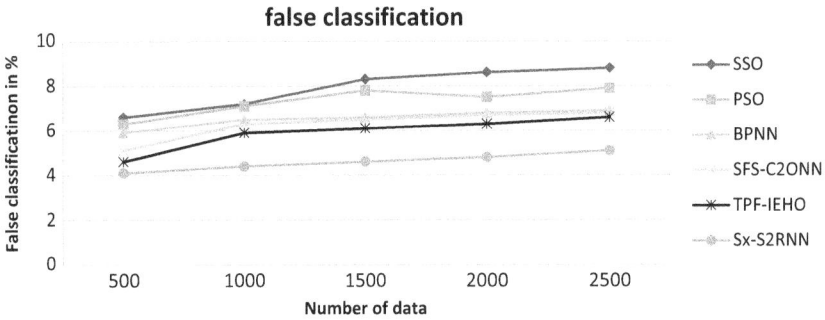

**FIGURE 9.6**
Performance of false classification.

**TABLE 9.5**

Performance of False Classification

| Methods/Datasets | Performance of False Classification in % | | | | | |
|---|---|---|---|---|---|---|
| | **SSO** | **PSO** | **BPNN** | **SFS-C²ONN** | **TPF-IEHO** | **Sx-S²RNN** |
| 500 | 6.6 | 6.3 | 5.9 | 5.1 | 4.6 | 4.1 |
| 1000 | 7.2 | 7.1 | 6.5 | 6.3 | 5.9 | 4.4 |
| 1500 | 8.3 | 7.8 | 6.6 | 6.5 | 6.1 | 4.6 |
| 2000 | 8.6 | 7.5 | 6.8 | 6.7 | 6.3 | 4.8 |
| 2500 | 8.8 | 7.9 | 6.9 | 6.8 | 6.6 | 5.1 |

dissimilar methods. The Sx-S²RNN attains lower false classification up to 4.1% than other evaluated methods of implementation.

$$\text{Time complexity}\left(\text{Tc}\right) = \sum_{k=0}^{k=n} \times \frac{\text{Total Features Handeled to Process in Dataset}}{\text{Time Taken}\left(\text{Ts}\right)} \qquad (9.16)$$

The detection accuracy is formulated to the time taken by a process. The detection accuracy is compared with different methods. The best detection takes the O (n) time by processing all the records based on the type definition of a defined category. The proposed system Sx-S²RNN produces a high 9.1 (ms) performance compared to all other previous systems, as shown in Figure 9.7.

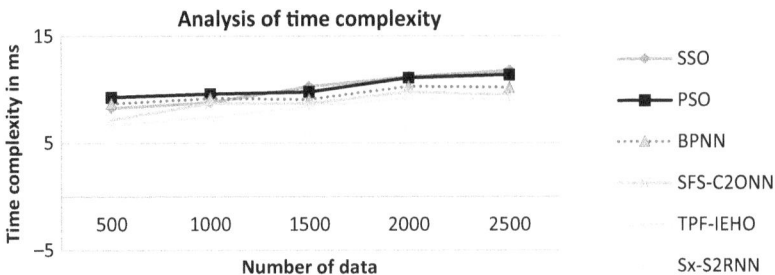

**FIGURE 9.7**
Performance of time complexity.

**TABLE 9.6**

Performance of Time Complexity

| Number of Data | Performance of Time Complexity in Milliseconds (ms) | | | | | |
|---|---|---|---|---|---|---|
| | SSO | PSO | BPNN | SFS-C²ONN | TPF-IEHO | Sx-S²RNN |
| 500 | 8.3 | 9.3 | 8.7 | 7.2 | 6.7 | 5.1 |
| 1000 | 8.8 | 9.6 | 9.2 | 8.7 | 7.4 | 5.4 |
| 1500 | 10.3 | 9.8 | 9.1 | 8.7 | 8.4 | 5.8 |
| 2000 | 11.2 | 11.1 | 10.3 | 9.8 | 9.2 | 6.1 |
| 2500 | 11.8 | 11.4 | 10.2 | 9.5 | 9.1 | 7.4 |

Table 9.6 indicates the total time taken to execute the process for predicting feature evaluation and classification depends on the feature limits at the finite state of N feature in O time. Table 9.6 shows the Performance of time evaluation in ms. The asymptotic O(n) is expressed to identify the processing time. The proposed system produces high performance with the least time taken up to 5.1 (ms) to execute the process than other methods.

## 9.5 Conclusion

To conclude, the DDoS type detection using phantom feature evaluation classifies results that prove the occurrence of widespread threats all over the network. The service intents need safety improvement by performing analysis of the feature selection and classification. In this proposed system, MS²F²A is important for detecting DDoS attacks. Based on this, lattice behavioral using Soft-Max Spread Spectral Recurrent Neural Network (Sx-S2RNN) has been used to detect DDoS attacks. Feature extraction models were built and trained using the trained level to provide input to the classification to get the best accuracy up to 98.1%. Traffic flow in-network behavioral conditions varies the features weight by considering service attack nature in the attacks. The proposed solution has proved the result based on classification methods producing high performance when compared to other methods. In future work, adaptive standard protocols can be implemented by reducing features that optimally give importance to an optimized deep neural network.

## References

1. S. Sumathi and N. Karthikeyan Detection of distributed denial of service using deep learning neural network. *J Ambient Intell Human Comput* (2020). doi:10.1007/s12652-020-02144-2.
2. L. Yang and H. Zhao, "DDoS attack identification and defense using SDN based on machine learning method," *Proceedings of the 15th International Symposium on Pervasive Systems, Algorithms and Networks (I-SPAN)*, IEEE, 2018, pp. 174–178.
3. M. Elsayed, A. LeKhac, S. Dev and A. Jurcut, "Machine-learning techniques for detecting attacks in SDN,"*Proceedings of the 7th International Conference on Computer Science and Network Technology*. IEEE, 2019.

4. R. Kokila, S. T. Selvi and K. Govindarajan, "DDoS detection and analysis in SDN-based environment using support vector machine classifier," *Proceedings of the Sixth International Conference on Advanced Computing (ICoAC)*. IEEE, 2014, pp. 205–221.

5. T. A. Tang, L. Mhamdi, D. McLernon, S. A. R. Zaidi and M. Ghogho, "Deep recurrent neural network for intrusion detection in SDN-based networks," *Proceedings of the 4th IEEE Conference on Network Softwarization and Workshops (NetSoft)*. IEEE, 2018, pp. 202–206.

6. J. Ye, X. Cheng, J. Zhu, L. Feng and L. Song, "A DDoS attack detection method based on SVM in a software-defined network," *Security and Communication Networks*, vol. 2018, 2018.

7. C. Yin, Y. Zhu, J. Fei and X. He, "A deep learning approach for intrusion detection using recurrent neural networks," *IEEE Access*, vol. 5, pp. 21954–21961, 2017.

8. F. Jiang, Y. Fu, B. B. Gupta, F. Lou, S. Rho, F. Meng and Z. Tian, "Deep learning based multi-channel intelligent attack detection for data security," *IEEE Transactions on Sustainable Computing*, vol. 5, pp. 204–212, 2018.

9. P. Jiangtao, C. Yunli and J. Wei, "A DDoS attack detection method based on machine learning." *Journal of Physics*, vol. 1237, p. 032040, 2019. doi:10.1088/1742-6596/1237/3/032040.

10. S. Das, D. Venugopal, S. Shiva and F. T. Sheldon, "Empirical evaluation of the ensemble framework for feature selection in DDoS attack" *International Conference on Edge Computing and Scalable Cloud (EdgeCom)*, 2020.

11. Y. Gu, K. Li and Z. Guo, Y. Wang, "Semi-supervised K-means DDoS detection method using hybrid feature selection algorithm," *IEEE Access*, vol. 7, pp. 64351–64365, 2019.

12. J. A. Pérez-Díaz, I. A. Valdovinos, K. -K. R. Choo and D. Zhu, "A flexible SDN-based architecture for identifying and mitigating low-rate DDoS attacks using machine learning," *IEEE Access*, vol. 8, pp. 155859–155872, 2020, doi: 10.1109/ACCESS.2020.3019330.

13. M. Tayyab, B. Belton and M. Anbar, "ICMPv6-Based DoS and DDoS attacks detection using machine learning techniques, open challenges, and blockchain applicability: A review," *IEEE Access*, vol. 8, pp. 170529–170547, 2020, doi: 10.1109/ACCESS.2020.3022963.

14. N. Zhang, F. Jaafar and Y. Malik, "Low-Rate DoS attack detection using PSD based entropy and machine learning," *6th IEEE International Conference on Cyber Security and Cloud Computing (CSCloud)/2019 5th IEEE International Conference on Edge Computing and Scalable Cloud (EdgeCom)*, Paris, France, 2019, pp. 59–62, doi: 10.1109/CSCloud/EdgeCom.2019.00020.

15. S.S. Priya, M. Sivaram, D. Yuvaraj and A. Jayanthiladevi, "Machine learning-based DDoS detection," *Tran. on Emerging Smart Computing and Informatics (ESCI)*, Pune, India, 2020, pp. 234–237, doi: 10.1109/ESCI48226.2020.9167642.

16. Z. He, T. Zhang and R. B. Lee, "Machine learning-based DDoS attack detection from source side in cloud," *IEEE 4th International Conference on Cyber Security and Cloud Computing (CSCloud)*, New York, NY, USA, 2017, pp. 114–120, doi: 10.1109/CSCloud.2017.58.

17. K. Wehbi, L. Hong, T. Al-Salah and A. A. Bhutta, "A survey on machine learning-based detection on DDoS attacks for IoT systems," *SoutheastCon*, Huntsville, AL, USA, 2019, pp. 1–6, doi: 10.1109/SoutheastCon42311.2019.9020468.

18. A. Bhati, A. Bouras, U. Ahmed Qidwai and A. Belhi, "Deep learning based identification of DDoS attacks in industrial application," *2020 Fourth World Conference on Smart Trends in Systems, Security and Sustainability (WorldS4)*, London, UK, 2020, pp. 190–196, doi: 10.1109/WorldS450073.2020.9210320.

19. M. Zekri, S. E. Kafhali, N. Aboutabit and Y. Saadi, "DDoS attack detection using machine learning techniques in cloud computing environments," *2017 Trans. on Cloud Computing Technologies and Applications (CloudTech)*, Rabat, Morocco, 2017, pp. 1–7, doi: 10.1109/CloudTech.2017.8284731.

20. M. Zekri, S. E. Kafhali, N. Aboutabit and Y. Saadi, "DDoS attack detection using machine learning techniques in cloud computing environments," *2017 3rd International Conference of Cloud Computing Technologies and Applications (CloudTech)*, Rabat, Morocco, 2017, pp. 1–7, doi: 10.1109/CloudTech.2017.8284731.

21. J. He, Y. Tan, W. Guo and M. Xian, "A small sample DDoS attack detection method based on deep transfer learning," *2020 International Conference on Computer Communication and Network Security (CCNS)*, Xi'an, China, 2020, pp. 47–50, doi: 10.1109/CCNS50731.2020.00019.
22. S. Yadav and S. Subramanian, "Detection of application layer DDoS attack by feature learning using Stacked AutoEncoder," *2016 Trans. Computational Techniques in Information and Communication Technologies (ICCTICT)*, New Delhi, India, 2016, pp. 361–366, doi: 10.1109/ICCTICT.2016.7514608.
23. P. Khuphiran, P. Leelaprute, P. Uthayopas, K. Ichikawa and W. Watanakeesuntorn, "Performance comparison of machine learning models for DDoS attacks detection," *2018 22nd International Computer Science and Engineering Conference (ICSEC)*, Chiang Mai, Thailand, 2018, pp. 1–4, doi: 10.1109/ICSEC.2018.8712757.

# 10

## Provably Secure Role Delegation Scheme for Medical Cyber-Physical Systems

**Rachana Y. Patil**

*Pimpri Chinchwad College of Engineering, Nigdi, Pune, India*

**Aparna Bannore**

*SIES Graduate School of Technology, Mumbai, India*

## CONTENTS

## 10.1  Introduction

Medical cyber-physical systems (MCPS) are the most life-critical system in the healthcare field. The growth of digitization and technology advancement has given rise to MCPS [1,2]. These days such systems are being used by majority of hospitals for providing a high quality of healthcare services.

DOI: 10.1201/9781003206088-10

The health and human service breach report says that the electronic health records (EHR) of patients are more important than credit card records in the darknet [3]. The existing healthcare systems are susceptible to data breaches, which can cause the compromise of sensitive EHR of patients.

Internet users in rural areas of countries like India have increased manifold in the past few years and are increasing exponentially. But relatively its appropriate application in some specific domains seems to be lagging, e.g., healthcare. This problem needs to be addressed by considering cyber-security threats [4] for healthcare data.

As a human right, the safety and security of patients' EHR is of prime importance and should not be compromised by any means. Maintaining data security on servers or on a cloud-based system is difficult, as service providers can misuse this sensitive EHR for any possible purpose [5,6]. Most of the time, casual human nature leads to continuing the same security keys for longer durations of days causing major EHR thefts.

The information produced by MCPS is not easily interoperable and available, and it becomes challenging for doctors to make decisions on patient treatment. The EHR, especially in medical imaging and pathological investigations, contains crucial and sensitive information that doctors use for detailed disease analysis [7]. When handling such data, internet use must therefore be highly protected.

Patients with chronic diseases require long-term treatment. In such cases, EHR from the cloud is accessed by medical experts, paramedical staff, and sometimes relatives of patients [8]. Due to the involvement of such multiple entities, data security is at high risk and it requires authorization at individual levels.

Based on the types of disease and its treatment, a specialist from the respective department will refer their opinion to another department. To provide secure data access in such cases, we propose the delegation-based forward-secure proxy signature scheme [9,10], which will provide data security at the individual level based on individual digital signature keys.

The architecture of MCPS is described in Figure 10.1. There are various sources of EHR such as data from any diagnostic instruments, implantable or wearable biosensors and pathological reports. This data is stored in the medical cloud for access by various entities related to the healthcare system. The hospitals need data for diagnosis of disease and deciding its line of treatment. Research institutions and the pharma industry will use it for research purposes and developing required drugs. The insurance sector needs EHR for verifying the authenticity of disease for claim settlement, while the disease control department will provide statistical analysis of specific diseases, their spread, and control.

### 10.1.1 Proxy Signature Technique

The basis of the idea is that, if the signing person is not available, he can appoint his/her signing capacitance [11–13]. Figure 10.2 shows the general process of applying a proxy signature. It works as follows.

1) The new authority receives a message to be delegated from the previous one, termed as the proxy signer.
2) Both the authorities, proxy and one who assigns, create the proxy key pair with a certain proxy algorithm and the public key is released.

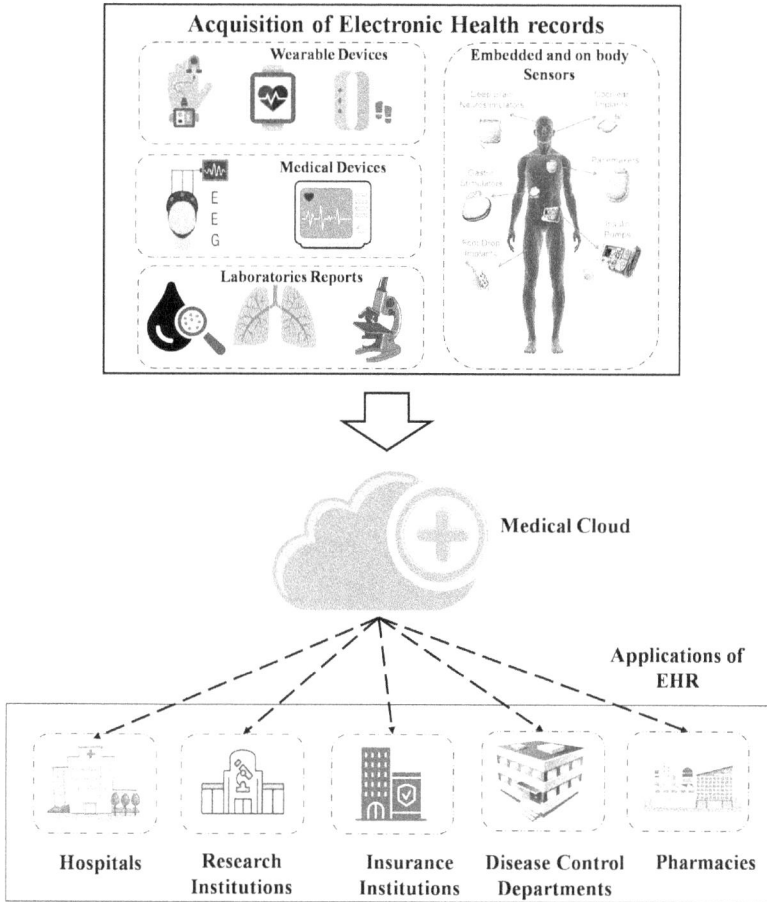

**FIGURE 10.1**
Architecture of MCPS.

3) Proxy authority applies a proxy secret key with the help of a digital signature algorithm.

4) The released public key is helpful for the receiver for signature validation.

## 10.1.2 Forward-Secure Technique

The concept of forwarding secrecy is advantageous in situations where the possibility of compromise of private is more [14,15].

The secrete key is altered after specific time intervals using cryptographic technics that assure the validity of signed papers. The public key remains unchanged for the complete time period of advanced secrecy, irrespective of updating the secret key after some time periods. All the documents/messages which are signed earlier will not change, except those which are signed during the secrete key leaking time period.

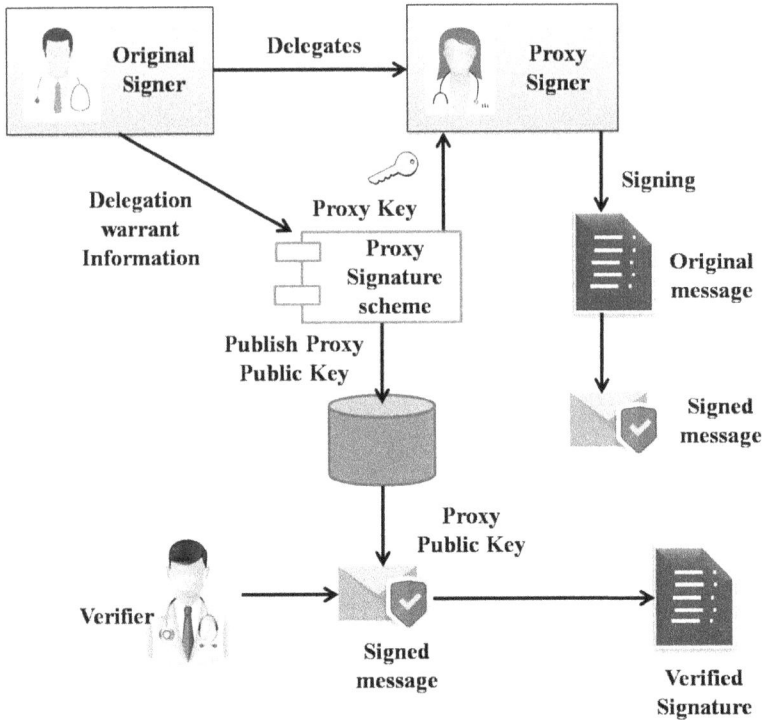

**FIGURE 10.2**
Proxy Signature Model.

Distinct proposals [16–19] for updating keys are discussed by the authors for maintaining and creating the secret key for the next time interval i+1. A general algorithm for updating the secret key is described underneath. The model of the forward-secure scheme is shown in Figure 10.3.

1) The original signer will be termed the initial pair of keys as (p_ra,p_rb).
2) Depending on the total period of forward secrecy, the forward-secure algorithm will calculate the public key for authentication.
3) The private key for the first period is calculated from the initial secrete key that is also treated as a seed secrete key.
4) The former secret key for a time period is utilized to sign in that time period.
5) For successive time periods, secrete key is calculated.
6) For advanced secrecy, the process of standard signature is suitably altered.

## 10.2 Related Work

In the literature, the cloud-based MCPS is proposed by authors of [8]. A trusted third party called an auditor can audit EHR without the need of retrieving the entire data.

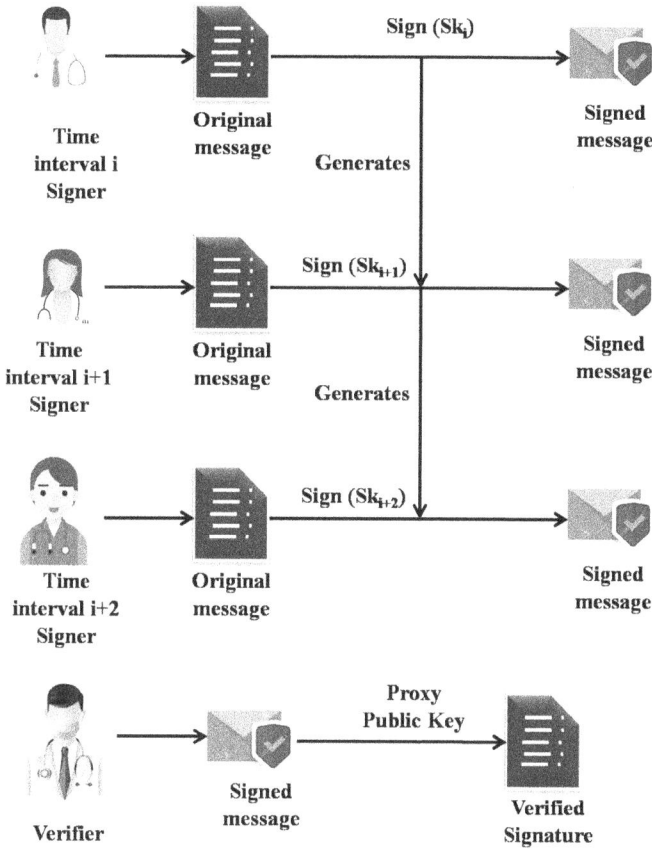

**FIGURE 10.3**
Forward Secure Signature.

The EHR are securely stored and shared through blockchain is proposed by authors of [21]. The certificate-less digital signature schemed is proposed for the purpose of authentication and integrity of EHR generated by medical equipment.

The mobile smartphone networks, which are widely used in MCPS, may behave maliciously. The trust-based Intrusion Detection System (IDS) for the network of such smartphones is used to identify the malicious nodes. The concept of Euclidean distance is used for identifying malicious Medical Smartphone Networks (MSN) nodes is based on behavioral profiles [6]. The authors of [5] have designed an $N^{th}$ degree Truncated polynomial Ring Units (NTRU) lattice based on an efficient and secure certificateless digital signature scheme that is resistant to quantum attack.

The lightweight scheme for wireless medical sensor networks is described in [21]. The proposed scheme is applied to cloud-based MCPS systems and is free from bilinear pairing.

The infrastructure for storing and processing is required by the MCPS applications that require delegation of roles.

Recently with blockchain technology growth of blockchain, many researchers have focused on developing healthcare systems using blockchains. On the other hand, these schemes [22,23] have not paid attention to the privacy protection requirement of MCPS.

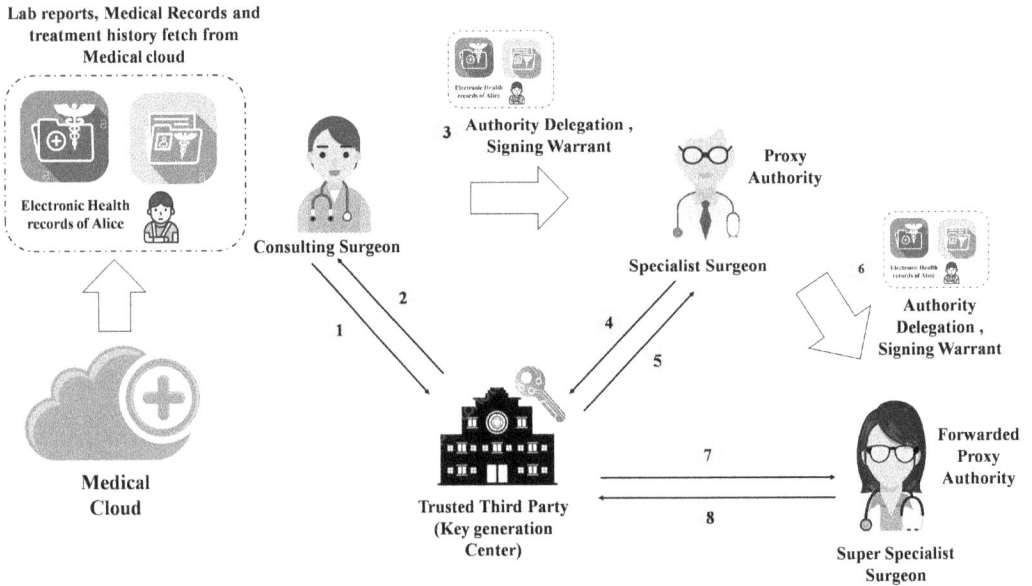

**FIGURE 10.4**
Medical cyber–Physical System with use of Proxy Signature.

The authors of [23] realized an efficient certificateless contract signing scheme for multi-party. The proposed system is computationally inefficient because of bilinear maps.

### 10.2.1 Case Study

As shown in Figure 10.4, the consulting surgeon wishes to securely provide the authorization of accessing and giving his opinion on the EHR of a patient (Alice), which further could be sent to the specialized doctors for a second opinion and deciding course of action.

In the above case, the consulting surgeon with access to Alice's EHR will prepare a proxy signature key pair and delegate his signing rights to the next specialist surgeon, which in turn will check the authenticity of the document received through warrant and provide his opinion and securely sign the data using his proxy signature key. Further, if any second opinion on the reports is required from the super specialty surgeon, then Alice's EHR are sent securely along with a signing warrant.

## 10.3 Framework for Proposed Methodology

As characterized in Figure 10.5, a delegation-based signature scheme for securely passing the records could be useful for maintaining the authorization security of the MCPS. The above-proposed system is developed using a pairing-based forward-secure proxy signature scheme (PB-FSPS) for securely authorizing the EHR that ensures the authenticity of the received records to the intermediaries and verifiers are able to easily audit the received data. In this proposed system the key generation and distribution required for authorization are handled by the trusted third party, thereby reducing the load on the MCPS system.

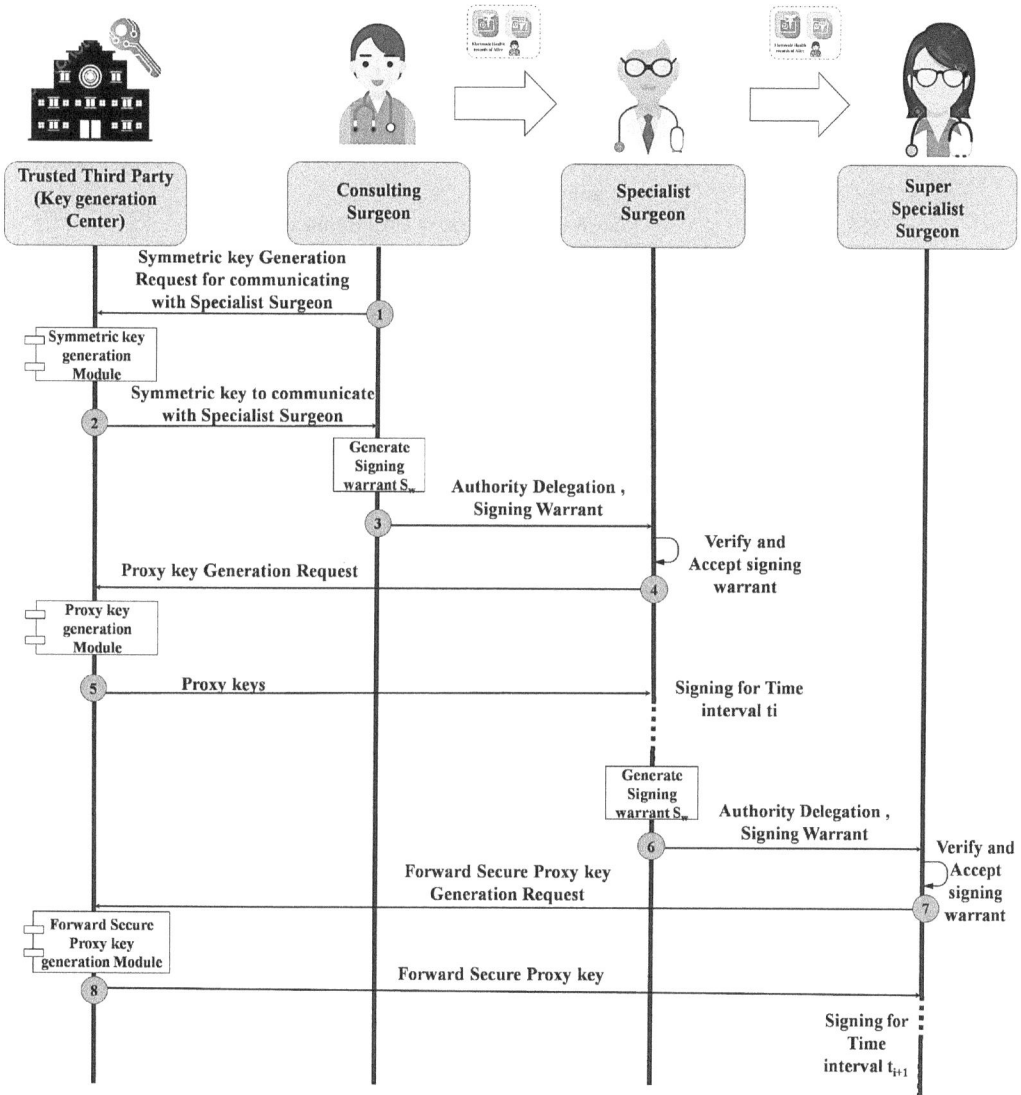

**FIGURE 10.5**
Sequence diagram representing flow of reports using proxy signature scheme.

## 10.3.1 The Proposed PB-FSPS Algorithm

The proposed PB-FSPS is discussed in this section. The algorithm consists of the following phases. Table 10.1 describes the notation used in the proposed algorithm.

### Setup Phase

Choose $<p, g, \mathbb{G}, \mathbb{G}_T, e>$
$\mathbb{G}, \mathbb{G}_T$: a cyclic additive and multiplicative group of prime order $p$, respectively
$g$ is a generator of $\mathbb{G}$
$e$: Bilinear map

**TABLE10.1**

Notations Using in Proposed Scheme

| | |
|---|---|
| $PK_o$, $PrK_o$ | Original signer public and private key, respectively |
| $PK_p$, $PrK_p$ | Proxy signer public and private key, respectively |
| $\mathcal{H}_1$, $\mathcal{H}_2$ | Hash functions |
| $S_w$ | Signing warrant |
| $FSK_i$ | Forward-secure proxy key for time interval $i$ |

Original signers keys ($PK_o$, $PrK_o$)
Proxy signers keys ($PK_p$, $PrK_p$)

### System Initialization/Define Public Parameters

$\mathcal{H}_1$: $\{0,1\}^* \to \mathbb{Z}_p*$
$\mathcal{H}_2$: $\{0,1\}^* \to \mathbb{G}$
e: $\mathbb{G} \times \mathbb{G} \to \mathbb{G}_T$
Common System Parameters are ($\mathbb{G}$, $\mathbb{G}_T$, $p$, $g$, $e$, $\mathcal{H}_1$, $\mathcal{H}_2$)

### Warrant Generation Phase

The warrant w is generated by the original signer that contains the information about the type of delegation and time of delegation, it also defines the type of documents to be signed by proxy signers.

By using warrant w, the original signer calculates signing warrant $S_w$ by using its own private key.

$$S_w = PrKo * \mathcal{H}_2(w)$$

Original signer delegates the signing rights by sending ($S_{w,\ w,} PK_o$).

### Warrant Verification Phase

The proxy signer verifies the received delegation by computing
$e(S_w, g) = e(\mathcal{H}_2(w), PK_o)$ if it is verified, then calculates

$$S'_w = S_w + PrK_p * \mathcal{H}_2(w)$$

### Forward-Secure Proxy Key Generation

This key is time-variant and valid only for time interval $t_i$

$$FSK_i = H_1\left(FSK_{i-1}^2\right)$$

The forward-secure public key for the entire time interval $T$

$$Q = H_1\left(FSK_0^{2^T}\right)$$

For the first time delegating the key is called as seed key.

The forward-secure private key for time interval 0 is $\text{FSK}_0 = S_w{}'$
The public key is $S_w{}'{*}g$

### Forward-Secure Proxy Signature Generation

M- Original message
Select the random number $r \rightarrow RZ_p*$
Calculate $u = e(g,g)^r$
Calculate $V = \mathcal{H}_1(M \parallel u)$
Calculate $S = V * \text{FSK}_i + r*g$
The proxy signer sends tuple $(V,S,M)$ to verifier

### Forward-Secure Proxy Signature Verification

A verifier can accept this forward-secure proxy signature if:

$$V = H_1\left(M \parallel e(S, g)\, e\left(H_2(w), Q\right)^{-V}\right)$$

### Correctness of Proposed Algorithm

Let us consider $e(S, g)\, e(H_2(w), Q)^{-V}$

$e(V * \text{FSK}_0 + r*g, g)\, e(H_2(w), Q)^{-V}$

$e(V * S_w' + r*g, g)\, e(H_2(w), Q)^{-V}$

$e(V * (S_{w+}PrK_p*g*H_2(w)) + r*g, g)\, e(H_2(w), Q)^{-V}$

$e(V*(S_{w+}PrK_p*g*H_2(w)))\, e(r*g, g)\, e(H_2(w), Q)^{-V}$

$e(V*(Q, H_2(w)))\, e(g,g)^r e(H_2(w), Q)^{-V}$

$e((H_2(w), Q))^V e(g,g)^r e(H_2(w), Q)^{-V}$

$e(g,g)^r = u$

$V = H_1(M \parallel e(S,g)\, e(H_2(w), Q)^{-V})$

$V = H_1(M \parallel u)$

## 10.4 Protocol Analysis Using BAN Logic

Figure 10.6 describes the process to be followed while analyzing the protocol using BAN logic [24,25]. The first protocol definition is written in BAN logic syntax in the Alice Bob notation. This is called idealization. In the assumption stage, the preliminary convention for the protocol has to be made clear. Initial expectations also indicate the communication parties' understanding. There should also be specific objectives that need to be evaluated. The BAN logic offers some previously established postulates or inference rules to prove whether the protocol meets and concludes the anticipated goal [26].

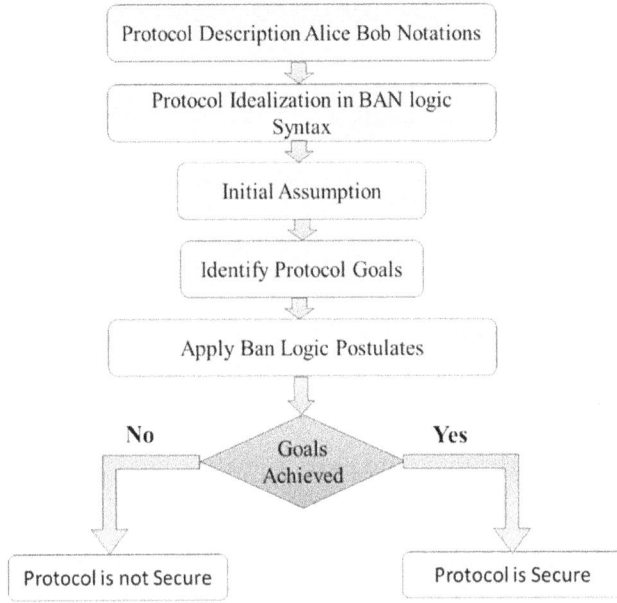

**FIGURE 10.6**
Process of BAN logic.

## 10.4.1 Idealized Protocol

Message 1: $O \rightarrow TTP : \{O| \sim ID_O, ID_A, N_1\}_{K_{TTP\_O}}$

Message 2: $TTP \rightarrow O : \left\{TTP| \sim O \xleftrightarrow{K_{O\_A}} A, ID_O, ID_A, N_1, \{O \xleftrightarrow{K_{O\_A}} A, ID_O, N_1\}_{K_{TTP\_A}}\right\}_{K_{TTP\_O}}$

Message 3: $O \rightarrow A : \{O| \sim O \xrightarrow{K_{O\_A}} A, ID_O, N_1\}_{K_{TTP\_A}}$

Message 4: $O \rightarrow A : \{O| \sim M_w, S_w, ID_O, M, U, | \xrightarrow{K_{O\_Pub}} O\}_{K_{O\_A}}$

Message 5: $A \rightarrow TTP : \{A| \sim S_w, N_2\}_{K_{TTP\_A}}$

Message 6: $TTP \rightarrow A : \left\{TTP | \sim | \xrightarrow{PK_{A\_Pub}} A, PK_{A\_Pub}^{-1}, N_2\right\}_{K_{TTP\_A}}$

Message 7: $A \rightarrow TTP : \{A| \sim ID_A, ID_B, N_3\}_{K_{TTP\_A}}$

Message 8: $TTP \rightarrow A \left\{TTP| \sim A \xleftrightarrow{K_{A\_B}} B, ID_B, N_3, \{ID_A, A \xleftrightarrow{K_{A\_B}} B, N_3\}_{K_{TTP\_B}}\right\}_{K_{TTP\_A}}$

Message 9: $A \rightarrow B : \{A| \sim ID_A, A \xleftrightarrow{K_{A\_B}} B, N_3\}_{K_{TTP\_B}}$

Message 10: $A \rightarrow B \{A| \sim M_w, M, U| \xrightarrow{K_{A\_Pub}} A\}_{K_{A\_B}}$

Message 11: $B \rightarrow TTP : \{B| \sim S_w, N_4\}_{K_{TTP\_B}}$

## 10.4.2 Initial Assumptions

For the derivation of proposed scheme using BAN logic, some of the initial assumptions are as follows:

A1: $O \mid \equiv \xrightarrow{K_{O\_Pub}} O$     A9: $\text{TTP} \mid \Rightarrow TTP \xleftarrow{K_{TTP\_O}} O$     A17: $A \mid \equiv A \xleftarrow{K_{A\_B}} B$

A2: $O \mid \equiv \xrightarrow{K_{O\_Pub}^{-1}} O$     A10: $\text{TTP} \mid \Rightarrow TTP \xleftarrow{K_{TTP\_A}} A$     A17: $B \mid \equiv \left( \text{TTP} \mid \Rightarrow A \xleftarrow{K_{A\_B}} B \right)$

A3: $A \mid \equiv \xrightarrow{K_{A\_Pub}} A$     A11: $\text{TTP} \mid \Rightarrow TTP \xleftarrow{K_{TTP\_B}} B$     A18: $O \mid \equiv \neq (N_1)$

A4: $A \mid \equiv \xrightarrow{K_{A\_Pub}^{-1}} A$     A12: $\text{TTP} \mid \Rightarrow O \xleftarrow{K_{O\_A}} A$     A19: $A \mid \equiv \neq (N_2)$

A5: $B \mid \equiv \xrightarrow{K_{B\_Pub}} B$     A13: $\text{TTP} \mid \Rightarrow A \xleftarrow{K_{A\_B}} B$     A20: $B \mid \equiv \neq (N_4)$

A6: $B \mid \equiv \xrightarrow{K_{B\_Pub}^{-1}} B$     A14: $\text{TTP} \mid \Rightarrow B \xleftarrow{K_{B\_C}} C$     A21: $A \mid \equiv TTP \xleftarrow{K_{TTP\_A}} A$

A7: $B \mid \equiv \xrightarrow{K_{A\_Pub}} A$     A15: $O \mid \equiv O \xleftarrow{K_{O\_A}} A$     A22: $\text{TTP} \mid \Rightarrow \mid \xrightarrow{PK_{A\_Pub}} A$

A8: $C \mid \equiv \xrightarrow{K_{B\_Pub}} B$     A16: $A \mid \equiv O \xleftarrow{K_{O\_A}} A$     A23: $TTP \mid \Rightarrow PK_{A_{\_Pub^{-1}}}$

## 10.4.3 Goals

The BAN logic is used to derive the following goals of the proposed scheme.

G1: $O \mid \equiv O \xleftarrow{K_{O\_A}} A$

G2: $A \mid \equiv O \xleftarrow{K_{O\_A}} A$

G3: $O \mid \equiv A \mid \equiv O \xleftarrow{K_{O\_A}} A$

G4: $A \mid \equiv O \mid \equiv O \xleftarrow{K_{O\_A}} A$

G5: $A \mid \equiv PK_{A\_Pub}$

G6: $A \mid \equiv PK_{A\_Pub}^{-1}$

## 10.4.4 Protocol Analysis

Idealized protocol messages are considered one after the other and established objectives are extracted from predefined deduction policy and hypotheses. The predetermined rules that can be used for the derivation of defined goals are described in detail in our previous work [26].

*Consider Message 2*

Message 2: $\text{TTP} \rightarrow O \left\{ TTP \mid \sim O \xleftarrow{K_{O\_A}} A, ID_O, ID_A, N_1, \left\{ O \xleftarrow{K_{O\_A}} A, ID_O, N_1 \right\}_{K_{TTP\_A}} \right\}_{K_{TTP\_O}}$

$O \triangleleft \left\{ TTP \mid \sim O \xleftarrow{K_{O\_A}} A, ID_O, ID_A, N_1, \left\{ O \xleftarrow{K_{O\_A}} A, ID_O, N_1 \right\}_{K_{TTP\_A}} \right\}_{K_{TTP\_O}}$

Apply MM 2 and A2

$$O| \equiv TTP| \sim O \xleftarrow{K_{O\_A}} A, ID_O, ID_A, N_1, \left\{ O \xleftarrow{K_{O\_A}} A, ID_O, N_1 \right\}_{K_{TTP\_A}}$$

$\left\{ O \xleftarrow{K_{O\_A}} A, ID_O, N_1 \right\}_{K_{TTP\_A}}$ is a Blackbox for original Signer O as $K_{TTP\_A}$ is a shared secret between TTP and A.

$$O \big| \equiv TTP \big| \sim O \xleftarrow{K_{O\_A}} A, ID_O, ID_A, N_1 \quad \text{Step 1}$$

Apply D2 and A18

$$O| \equiv \neq \left( O \xleftarrow{K_{O\_A}} A, ID_O, ID_A, N_1 \right) \quad \text{Step 2}$$

Apply nonce verification (NV) and use Steps 1 and 2

$$O| \equiv TTP| \equiv O \xleftarrow{K_{O\_A}} A, ID_O, ID_A, N_1$$

Apply belief decomposition rule (D3)

$$O| \equiv TTP| \equiv O \xleftarrow{K_{O\_A}} A$$

Apply jurisdiction rule and assumption (A12)

$$O| \equiv O \xleftarrow{K_{O\_A}} A \quad \text{Goal 1}$$

**Consider Message 3**

Message 3: $O \rightarrow A : \left\{ O| \sim O \xleftarrow{K_{O\_A}} A, ID_O, N_1 \right\}_{K_{TTP\_A}}$

$$A \triangleleft \left\{ O| \sim O \xleftarrow{K_{O\_A}} A, ID_O, N_1 \right\}_{K_{TTP\_A}}$$

Apply MM 2 and A21

$$A| \equiv O| \sim O \xleftarrow{K_{O\_A}} A, ID_O, N_1 \quad \text{Step 3}$$

Apply FD2 and A18

$$A| \equiv \neq \left( O \xleftarrow{K_{O\_A}} A, ID_O, N_1 \right) \quad \text{Step 4}$$

Apply NV and use Steps 1 and 2

$$A| \equiv O| \equiv O \xleftarrow{\ K_{O\_A}\ } A, ID_O, N_1$$

Apply belief decomposition rule (BD3)

$$A| \equiv O| \equiv O \xleftarrow{\ K_{O\_A}\ } A \qquad \text{Goal 4}$$

Apply jurisdiction rule

$$A| \equiv O \xleftarrow{\ K_{OA}\ } A \qquad \text{Goal 2}$$

**Consider Message 6**

Message 6: $TTP \to A : \left\{ TTP| \sim | \xleftarrow{\ PK_{A\_Pub}\ } A, \quad PK_{A\_Pub}^{-1}, N_2 \right\}_{K_{TTP\_A}}$

$$A \triangleleft \left\{ TTP| \sim | \xleftarrow{\ PK_{A\_pub}\ } A, \quad PK_{A\_pub}^{-1}, N_2 \right\}_{K_{TTP\_A}}$$

Apply MM 2 and A21

$$A| \equiv TTP| \sim | \xrightarrow{\ PK_{APub}\ } A, \quad PK_{APub}^{-1}, N_2$$

Apply FD2 and A19

$$A| \equiv \neq \left( | \xrightarrow{\ PK_{APub}\ } A, PK_{APub}^{-1}, N_2 \right)$$

Apply NV:

$$A| \equiv TTP | \equiv | \xrightarrow{\ PK_{APub}\ } A, \quad PK_{APub}^{-1}, N_2$$

Apply belief decomposition rule (BD3)

$$A| \equiv TTP | \equiv | \xrightarrow{\ PK_{A\_Pub}\ } A, \quad PK_{A\_Pub}^{-1} \qquad \text{Step 5}$$

Apply jurisdiction rule and assumption (A22) to step 5

$$A| \equiv TTP| \equiv | \xrightarrow{\ PK_{A\_Pub}\ } A \qquad \text{Goal 5}$$

Apply jurisdiction rule and assumption (A22) to Step 5

$$A | \equiv PK_{APub}^{-1} \qquad \text{Goal 6}$$

## 10.5 Simulation Study of PB-FSPS

AVISPA is used to simulate the proposed forward-secure proxy signature scheme's security proof [27,28]. The findings show that PB-FSPS is tolerant to replay and Man in the Middle (MITM) attacks. It should be emphasized that AVISPA exclusively tackles replay and MITM threats for any security protocol.

The basic role of the participating entities written in High-Level Protocol Specification Language (HLPSL) [29,30] in a protocol defines some sequence of actions. Later, the agents

```
%%HLPSL:

role OriginalSigner (OS, DA, DB , TTP :agent,
                     Hash : hash_func,
                     Ko, Ka, Kb, Kc0,Kc1 : public_key,
                                          Kot, Kat, Kbt, Koa, Kab : symmetric_key,
                     M : message,
                     SND, RCV: channel (dy))

played_by OS def=

local State : nat,
      M, Mw, Sw, n0, U, IDo, IDa, N1, N2 : text,
         Kot, Kat: symmetric_key

             init State := 0

             transition

      0.   State  = 0 /\ RCV(start) =|>
      State':= 2 /\ N1'  := new()
                          /\ SND(N1'.IDa.IDb)


      2.   State =2 /\ RCV({Koa.IDo.IDa.N1'.{Koa.IDo.N1'}_Kat}_Kot) =|>
                State':=4 /\ SND({Koa.IDo.N1'}_Kat)
                                /\ Mw:= new() /\ Sw:=new() /\ n0:=new() /\ U=new()
                                /\ SND({Mw,Sw,Ko,U}_Koa)
                                /\ secret(N1',n1,{OS,DA})
                        /\ witness(OS,DA,OriginalSigner_DelegeteeA_n1')


             end role
```

**FIGURE 10.7**
The role specification for original signer.

```
role TTP (OS, DA, DB , TTP :agent,
                    Hash : hash_func,
                    Ko, Ka, Kb, Kc0,Kc1 : public_key,
                                         Kot, Kat, Kbt, Koa, Kab : symmetric_key,
                    M : message,
                    SND, RCV: channel (dy))
played_by TTP def=

local State : nat,
      M, Mw, Sw, U, IDo, IDa, N1, N2 : text,
            Kc0, Kc1 : public_key,
         Kot, Kat: symmetric_key

               init State := 1

               transition

      1.  State  = 1 /\ RCV(N1'.IDa.IDb) =|>
      State':= 3 /\ Koa := new()
                        /\ SND(Koa.IDo.IDa.N1'.{Koa.IDo.N1'}_Kat}_Kot)

      3.  State =    3 /\RCV({Sw.N2}_Kat)=|>
            State':=9 /\Kc :=new() /\ inv(Kc):=new
                            /\SND({Kc0.inv(Kc0).N2'}_Kat)

      9. State=9         /\ RCV({IDa.IDb.N3}_Kat) =|>
      State':=13 /\ SND(Kab.IDa.IDb.N3'.{Kab.IDa.N3'}_Kbt}_Kat)

      13. State=13   /\RCV({Sw.N4}_Kbt) =|>
            State':=21 /\Kc1 :=new() /\ inv(Kc1):=new
                            /\SND({Kc1.inv(Kc1).N4'}_Kbt)

      end role
```

**FIGURE 10.8**
The role specification for trusted third party (TTP).

instantiate this condition. The four primary roles in the projected forward-secure proxy signature protocol, which we call original signer and trusted third party, delegatee A and delegatee B.

The role specification for the original signer is described in Figure 10.7. original signer is originally in state 0 after receiving the start message the state of the original signer changes from 0 to 1 and sends the identity of the proxy signer (delegetee A) and his own identity O to a trusted third party securely. The Dolev–Yao intruder model is used for implementing. Snd( ) and.Rcv( ) channels. The role specification of TTP is described in Figure 10.8.

```
role DelegeteeA (OS, DA, DB , TTP :agent,
                  Hash : hash_func,
                  Ko, Ka, Kb, Kc0,Kc1 : public_key,
                                   Kat, Koa : symmetric_key,
                  M : message,
                  SND, RCV: channel (dy))

        played_by DA def=
local State : nat,
        M, Mw, Sw, n0, U, IDo, IDa, IDb, N1, N2, N3 : text,
             Kc,Ko,Ka : public_key,
          Kot, Kat: symmetric_key

             init State := 5

             transition



   5.   State =5   /\ RCV({Koa.IDo.N1'}_Kat)
                    /\ RCV({Mw,Sw,Ko,U}_Koa)=|>
           State':=7 /\ N2' := new()
                               /\SND({Sw.N2}_Kat)

   7.       State =7   /\ RCV({Kc.inv(Kc).N2'}_Kat)=|>
          State':=11 /\ N3':=new()
                               /\ SND({IDa.IDb.N3}_Kat)

   11. State =11 /\ RCV(Kab.IDa.IDb.N3'.{Kab.IDa.N3'}_Kbt}_Kat)
        State':=15 /\SND({Kab.IDa.N3'}_Kbt)
                           /\ Mw:= new() /\ Sw:=new() /\ n0:=new() /\ U=new()
                           /\ SND({Mw,Sw,Ka,U}_Kab)
                           /\ secret(N3',n3,{DA,DB})
                      /\ witness(DA,DB,DelegeteeA_DelegeteeB_n3')



      end role
```

**FIGURE 10.9**
The role specification for delegatee A (DA).


Likewise, the delegatee A and delegatee B roles are specified as shown in Figures 10.9 and 10.10, respectively. The declaration witness (DA,DB,delegeteea_delegeteeb_n3, N3') signifies that a nonce N3 produced by delegatee A is fresh for delegetee B and request(DA,DB,delegeteea_delegeteeb_n3,N3') represents delegetee B's reception of a nonce N3 that was generated by delegetee A for delegetee B. The declaration secret(N3',n3,{DA,DB}) ensure that the generated nonce N3 is kept secret among delegetee A and delegetee B.

Figures 10.11 and 10.12 specify the session and environment roles, respectively. When the OFMC backend is launched with the default settings, it can be seen that OFMC has found no attacks. For restricted sessions, the BP-FSPS is also implemented with the CL-AtSe backend. The output in Figure 10.13 demonstrates that the protocol is likewise safe while using CL-AtSe.

```
played_by DB def=

local State : nat,
        M, Mw, Sw, n0, U, IDo, IDa, IDb, N1, N2, N3, N4 : text,
            Kc,Ko,Ka : public_key,
        Kat, Kbt: symmetric_key

                init State := 17

                transition

    17.  State =17  /\ RCV({Kab.IDa.N3'}_Kbt)
                /\ RCV({Mw,Sw,Ka,U}_Kab)=|>
            State':=19 /\ N4' := new()
                                /\SND({Sw.N4}_Kbt)

    19. State=19 /\ RCV({Kc1.inv(Kc1).N4'}_Kbt)


        end role
```

**FIGURE 10.10**
The role specification for delegetee B (DB).

```
role session(OS, DA, DB , TTP :agent,
                    Ko, Ka, Kb, Kc0,Kc1 : public_key,
                    Kot, Kat, Kbt, Koa, Kab : symmetric_key,
                    Hash : hash_func,
                    )
def=
local SO, RO, SA, RA, SB ,RB, ST, RT: channel(dy)

composition

        OriginalSigner(OS,DA,DB,TTP,Ko,Ka,Kb,Kot,Koa,Kc0,SO,RO)
        /\
        TTP(OS,DA,DB,TTP,Ko,Ka,Kb,Kot,Koa,Kab,Kc0,Kc1,St,Rt)
        /\
        DelegeteeA(OS,DA,DB,TTP,Ko,Ka,Kb,Kat,Koa,Kab,Kc0,SA,RA)
        /\
        DelegeteeB(OS,DA,DB,TTP,Ko,Ka,Kb,Kbt,Kab,Kc1,SB,RB)

end role
```

**FIGURE 10.11**
Session role.

```
role environment()
def=
        const
                os,da,db,ttp :agent,
                ko, ka, kb, kc0,kc1 : public_key,
                kot, kat, kbt, koa, kab, Koi, Kai : symmetric_key,
                hash : hash_func,
                n1,n2,n3,n4,
                OriginalSigner_DelegeteeA_n1,
                DelegeteeA_DelegeteeB_n3:protocol_id


    intruder_knowledge = {os,da,db,ttp,ko, ka, kb}

    composition

        session(os,ttp,da,ko,Ka,kot,kat,koa)
     /\ session(os,ttp,i,ko,Ki,kot.kit,koi)
     /\ session(i,ttp,da,ki,Ka,kit.kat,koi)

end role

%%%%%%%%%%%%%%%%%%%%%%%%%%%%%%%%%%%%%%%%%%%%%%%%%%%%%%%%%%%%%%%%
goal
  secrecy_of n1,n2,n3,n4,koa,kab
  authentication_on OriginalSigner_DelegeteeA_n1
  authentication_on DelegeteeA_DelegeteeB_n3

end goal
%%%%%%%%%%%%%%%%%%%%%%%%%%%%%%%%%%%%%%%%%%%%%%%%%%%%%%%%%%%%%%%%
environment()
```

**FIGURE 10.12**
Environment role.

| | |
|---|---|
| % OFMC | SUMMARY |
| % Version of 2006/02/13 | SAFE |
| SUMMARY | DETAILS |
| SAFE | BOUNDED_NUMBER_OF_SESSIONS |
| DETAILS | TYPED_MODEL |
| BOUNDED_NUMBER_OF_SESSIONS | PROTOCOL |
| PROTOCOL | /home/span/span/testsuite/results/MCPSV3.if |
| /home/span/span/testsuite/results/MCPSV3.if | GOAL |
| GOAL | As Specified |
| as_specified | BACKEND |
| BACKEND | CL-AtSe |
| OFMC | STATISTICS |
| COMMENTS | Analysed :19 states |
| STATISTICS | Reachable :10 states |
| parseTime: 0.00s | Translation: 0.14 seconds |
| searchTime: 1.15s | Computation: 0.00 seconds |
| visitedNodes: 347 nodes | |
| depth: 13 plies | |

**FIGURE 10.13**
AVISPA Output.

## 10.6 Conclusion

One of the powerful gifts of humanity was the use of technology in the area of critical disease treatment. The virtual world is joined to the real one by the MCPS. In addition to MCPS facilities, however, the compromised EHR of patients is the biggest concern. The protection and safety of these records are a new challenge and a key priority. Cyber-security attacks on healthcare facilities affect human lives directly. EHR are essential to a security concern for their privacy and authenticity. In this chapter, we proposed the bilinear map-based EHR delegation system BP-FSPS scheme. In case of critical illness, the patient must be treated by several specialists, with a treatment line that matches one another. Without violating confidentiality, delegations may have safe access to the EHRs in these circumstances.

Delegation-based authentication and authorization are provided by proxy signature systems. The proposed scheme is verified through mathematical modeling, formal verification, and simulation.

## References

1. Qiu, H., Qiu, M., Liu, M., & Memmi, G. (2020). Secure health data sharing for medical cyber-physical systems for the healthcare 4.0. *IEEE Journal of Biomedical and Health Informatics*, 24(9), 2499–2505.
2. Xu, Z., He, D., Wang, H., Vijayakumar, P., & Choo, K. K. R. (2020). A novel proxy-oriented public auditing scheme for cloud-based medical cyber physical systems. *Journal of Information Security and Applications*, 51, 102453.
3. Dey, N., Ashour, A. S., Shi, F., Fong, S. J., & Tavares, J. M. R. (2018). Medical cyber-physical systems: A survey. *Journal of Medical Systems*, 42(4), 1–13.
4. Patil, R. Y., & Devane, S. R. (2017, October). Unmasking of source identity, a step beyond in cyber forensic. In *Proceedings of the 10th International Conference on Security of Information and Networks* (pp. 157–164).
5. Xu, Z., He, D., Vijayakumar, P., Choo, K. K. R., & Li, L. (2020). Efficient NTRU lattice-based certificateless signature scheme for medical cyber-physical systems. *Journal of Medical Systems*, 44(5), 1–8.
6. Meng, W., Li, W., Wang, Y., & Au, M. H. (2020). Detecting insider attacks in medical cyber-physical networks based on behavioral profiling. *Future Generation Computer Systems*, 108, 1258–1266.
7. Shu, H., Qi, P., Huang, Y., Chen, F., Xie, D., & Sun, L. (2020). An efficient certificateless aggregate signature scheme for blockchain-based medical cyber physical systems. *Sensors*, 20(5), 1521.
8. Zhang, X., Zhao, J., Mu, L., Tang, Y., & Xu, C. (2019). Identity-based proxy-oriented outsourcing with public auditing in cloud-based medical cyber-physical systems. *Pervasive and Mobile Computing*, 56, 18–28.
9. Sunitha, N. R., & Amberker, B. B. (2008). Proxy signature schemes for controlled delegation. *Journal of Information Assurance and Security*, 3(2), 159–174.
10. Sunitha, N. R., Amberker, B. B., & Koulgi, P. (2007, October). Controlled delegation in e-cheques using proxy signatures. In *11th IEEE International Enterprise Distributed Object Computing Conference (EDOC 2007, Kunming, China)* (pp. 414–414). IEEE.
11. Lee, B., Kim, H., & Kim, K. (2001, January). Strong proxy signature and its applications. In *Proceedings of SCIS* (Vol. 2001, pp. 603–608).
12. Boldyreva, A., Palacio, A., & Warinschi, B. (2012). Secure proxy signature schemes for delegation of signing rights. *Journal of Cryptology*, 25(1), 57–115.
13. Sun, H. M. (1999). An efficient nonrepudiable threshold proxy signature scheme with known signers. *Computer Communications*, 22(8), 717–722.
14. Amberker, B. B., Koulgi, P., & Sunitha, N. R. (2007, April). Forward-Security for ElGamal-like Signature Scheme. In *Proceedings of 6th Annual Security Conference*, Las Vegas.
15. Yali, L., Xinchun, Y., Chunlai, S., & Hailing, Z. (2008, July). A new forward-secure signature scheme. In *2008 27th Chinese Control Conference, Kunming, China* (pp. 821–825). IEEE.
16. Jun, H., Ximei, L., Lijuan, L., & Chunming, T. A New Forward-Secure Proxy Signature Scheme. In *2010 International Forum on Information Technology and Applications, Kunming, China.*
17. Abdelfatah, R. I. (2017). A novel proxy signcryption scheme and its elliptic curve variant. *International Journal of Computer Applications*, 165(2), 36–43.
18. Dutta, M., Singh, A. K., & Kumar, A. (2013, February). An efficient signcryption scheme based on ECC with forward secrecy and encrypted message authentication. In *2013 3rd IEEE International Advance Computing Conference (IACC)* (pp. 399–403). IEEE.

19. Fazlagic, S., Behlilovic, N., & Mrdovic, S. (2012, November). Controlled delegation of signature in workflow systems. In *2012 20th Telecommunications Forum (TELFOR)* (pp. 1389–1392). IEEE.

20. Bhole, D., Aditi, M., and Rachana, P. (2016, April). A new security protocol using hybrid cryptography algorithms. *International Journal of Computer Sciences and Engineering*4(2), 18–22.

21. Zhang, X., Xu, C., Zhang, Y., & Jin, C. (2017). Efficient integrity verification scheme for medical data records in cloud-assisted wireless medical sensor networks. *Wireless Personal Communications*, 96(2), 1819–1833.

22. Suzhen, C. A. O., Fei, W. A. N. G., Xiaoli, L. A. N. G., Rui, W. A. N. G., & Xueyan, L. I. U. (2019). Multi-party contract signing protocol based on certificateless.*41*(11), 2691–2698.

23. Zhao, Y. (2018). Aggregation of gamma-signatures and applications to bitcoin. *IACR Cryptol. ePrint Arch., 2018*, 414.

24. Burrows, M., Abadi, M., & Needham, R. M. (1989). A logic of authentication. *Proceedings of the Royal Society of London A: Mathematical and Physical Sciences*, 426(1871), 233–271.

25. Teepe, W. (2009). On BAN logic and hash functions or: How an unjustified inference rule causes problems. *Autonomous Agents and Multi-Agent Systems*, 19(1), 76–88.

26. Yogesh, P. R. (2020). Formal verification of secure evidence collection protocol using BAN logic and AVISPA. *Procedia Computer Science*, 167, 1334–1344.

27. Patil, R. Y., & Devane, S. R. (2019). Network forensic investigation protocol to identify true origin of cyber crime. *Journal of King Saud University-Computer and Information Sciences*. Doi: 10.1016/j.jksuci.2019.11.016.

28. Armando, A., Basin, D., Boichut, Y., Chevalier, Y., Compagna, L., Cuéllar, J.,… & Vigneron, L. (2005, July). The AVISPA tool for the automated validation of internet security protocols and applications. In *International Conference on Computer Aided Verification* (pp. 281–285). Springer, Berlin, Heidelberg.

29. Von Oheimb, D. (2005, September). The high-level protocol specification language HLPSL developed in the EU project AVISPA. In *Proceedings of APPSEM 2005 workshop* (pp. 1–17).

30. Glouche, Y., Genet, T., Heen, O., & Courtay, O. (2006, May). A security protocol animator tool for AVISPA. In *ARTIST2 Workshop on Security Specification and Verification of Embedded Systems*, Pisa.

# 11

## FRAME Routing to Handle Security Attacks in Wireless Sensor Networks

**Subramanyam Radha**

*Karunya Institute of Technology and Sciences, CBE, India*

**Bidar Sachin and Seyedali Pourmoafi**

*University of Hertfordshire, Hatfield, United Kingdom*

**Perattur Nagabushanam**

*Karunya Institute of Technology and Sciences, CBE, India*

## CONTENTS

## 11.1 Introduction

Wireless sensor network (WSN) helps to estimate physical parameters of interest in noisy measurements through the cooperation among sensor nodes. WSN performs estimated task distributions based on the local measurements, computation by each node, information exchange with neighboring nodes. Compared with centralized estimation, advantages of distributed estimation (i) overhead of transmitting all data to fusion center can be reduced; (ii) the estimation is resilient, robust against sensor failures. Distributed

DOI: 10.1201/9781003206088-11

estimation methods are developed with applications in many fields, such as environmental monitoring, power allocation and industrial automation.

In distributed estimation, the sensors are required to estimate a few parameters to improve accuracy. Most of the existing research investigates homogeneous networks, all sensors possess identical communication, computation capability and their roles in distributed estimation are homogeneous. It is reported that heterogeneity in sensors of a network can prolong the network's lifetime, improve communication reliability. Now, the question is whether it is better for a distributed estimation to use the combination of different nodes; for example, a large number of cheap, low-end nodes for communication and simple data aggregation, combined with few high-end nodes for executing powerful functions in terms of sensing, communication, and computation. The effectiveness of heterogeneous sensor networks which consist of sensor nodes and relay nodes has been analyzed in this work.

Single-tier, two-tier WSN is implemented with delay constraint in relay nodes and with a set covering algorithm which forms a tree from source to sink node to transmit packets. Delay in relay node placement is a challenging task; however, we have used a two-tier approach for connectivity aware coverage to overcome it. Connectivity in disjoint networks is a major issue, which is addressed here using convex hull polygon and zero gradient point inside it. Fidelity and distortion constraints can be addressed using a relay model with Gaussian degradation. Relay nodes are separated from cluster heads in WSN using distributive clustering with mutual exclusive logic. Clustering helps in reducing communication energy; however, if a cluster head acts as a relay, then aggregation and multi-hop communication lead to problems. Assisted beam-forming and adaptive beam-forming help to reduce path loss which limits transmission distance. Beam-forming directional antennas are used for both transmission and reception of packets, which in turn helps in overcoming path loss.

The objectives in this work are

- To find an energy-efficient path from source to destination using FRAME routing protocol (network layer). And to analyze the performance of WSN in FRAME routing.
- To improve WSN network performance in terms of energy consumption, packets dropping ratio and throughput.

The overview of the work carried out is shown in Figure 11.1. Conventional routing and FRAME routing are analyzed in with and without attacks environment. In Section 11.2, a detailed literature survey is carried out on approaches to detect attacks in WSN, Internet of Things (IoT), and other low power systems and architectures.

Ad-Hoc On-Demand Vector (AODV) for ad hoc networks and FRAME routing with Bayesian network for wireless networks is described in Section 11.3. In Section 11.4, the respective results of AODV and FRAME are shown. Section 11.5 concludes the work by proposing a few future directions.

## 11.2 Literature Survey

Active attacks like a black hole, warm hole attacks, and passive attacks like homing attacks and eavesdropping are analyzed with false messages and traffic analysis by Keerthika and Shanmugapriya (2021). Various countermeasures to save the WSN network from

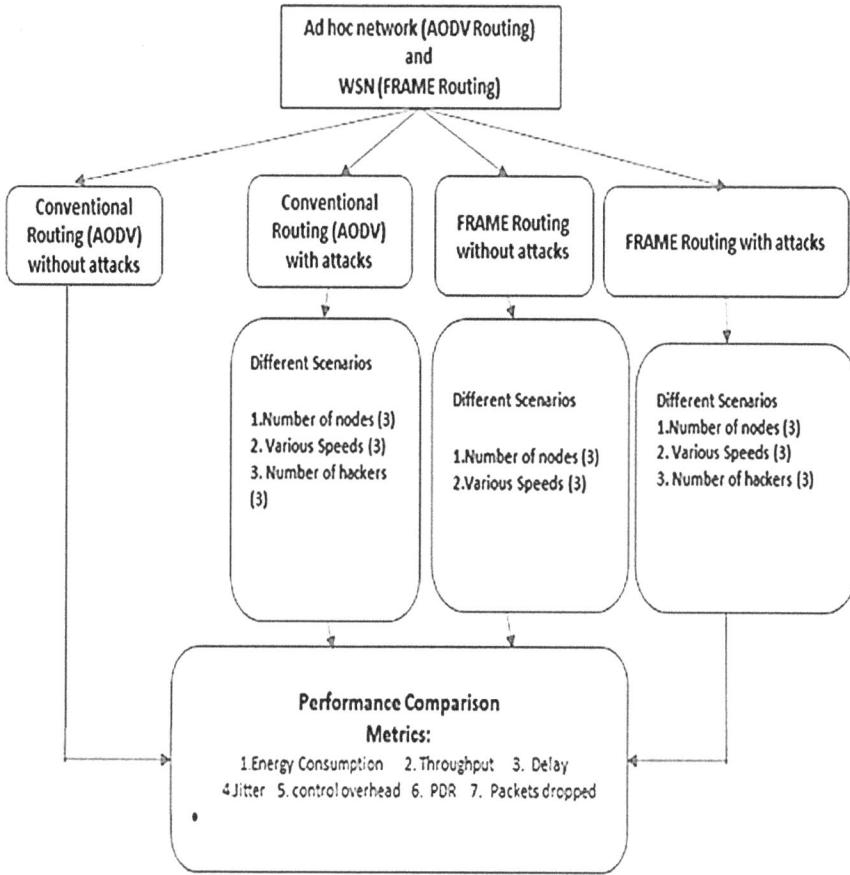

**FIGURE 11.1**
Overview of the work.

such active and passive attacks are proposed and analyzed to improve network performance. Node privacy in WSN is analyzed in terms of energy analysis with the help of tinyOS and test-bed implementation. Elliptical curve cryptography algorithm is used for addressing security and functional requirements in the network by Ummer Iqbal et al. (2020). The trusted authentication protocol is used to address security and privacy in WSN by Yuan Gao et al. (2018). Mica2 and crossbow motes are used for implementation, and security enhancements and fidelity are achieved. However, the authors did not present the performance numerically or graphically of the network tested by them with their approaches.

Anonymous access (AAA) authenticate algorithm is used to address security issues in WSN for the Big Data environment by Shadi Nashwan (2021). High security and efficiency is achieved for the network and discussed in terms of security and computational cost analysis. Chander and Kumaravelan (2021) made a detailed survey of ML and AI approaches for detecting abnormalities, and anomalies/outliers in WSN. The approaches are analyzed in terms of computational capabilities and minimum power resources usage. Different layers like network layer, perception layer and the attacks like jamming, collision, and exhaustion attacks for IoT, RFID are analyzed by Huanan et al. (2021). Countermeasures like cryptography, error-correcting codes, rate limiting, raising alarm are proposed by the authors. Optimization techniques for node security and fusion for data security in WSN are concentrated by Huanan et al. (2021). Secret key management is generated to ensure low computational complexity, storage, and scalability. Abnormal sensors detection accuracy (ASDA), RSA and interlock algorithms are proposed to prevent WSN from DoS attacks by Fotohi et al. (2020). Throughput and PDR are analyzed using NS2 tool and to improve the performance of WSN. Table 11.1 shows literature on attacks for WSN.

Security attacks in short-range wireless communication like Bluetooth, wi-fi etc. are analyzed and countermeasures are proposed by Lounis et al. (2020). Reducing anonymity, transmission power and out-of-band association can be followed to have strong authentication in wireless technologies and IoT applications. Software-defined radio (SDN) and security requirements are addressed for IoT applications by Iqbal et al. (2020). General security issues, data security issues are analyzed and unifying stakeholders helps in better performance of the network. Link flooding, DDoS attacks in IoT are analyzed and

**TABLE 11.1**

Literature Survey for Attacks in WSN

| Authors (year) | Algorithm/Technique | Merits and Demerits | Software and Metrics Used | Applications |
|---|---|---|---|---|
| Keerthika et al. (2021) | Spoof routing, selective forwarding | Counter measures for active and passive attacks | False messages, traffic analysis | WSN |
| Gao et al. (2018) | Trusted authentication protocol | Security enhancement, fidelity | Mica2, crossbow motes | WSN |
| Huanan et al. (2021) | Optimization for node security, fusion for data security | Secret key management | Storage, computational capability | WSN |
| Fotohi et al. (2020) | Cryptography RSA algorithm | Authentication methods, interlock protocol to prevent DoS | NS2, PDR and throughput | WSN |
| Abidoye and Obagbuwa (2018) | Coordinated attacks by bogus requests | Counter measures for DDoS | Detect, defend the DDoS attacks | WSN |
| Anitha et al. (2021) | Ant colony optimization, moving average for replica detection | Higher detection probability for healthcare monitoring | Storage, control overheads | WSN |
| Bhatt et al. (2020) | Genetic, fruit fly optimization algorithms | Least resource, maximum node contribution | Energy cost and efficiency | WSN |
| Hongsong et al. (2020) | Hilbert huang transform (HTT) for denial of service | Attack detection using correlation analysis | Pearson, spearman coefficients | WSN |
| Fu et al. (2018) | Power distribution in cyber physical systems | Graph theory to detect firewalls and intrusions | Throughput, robustness, security | WSN |
| Wazid et al. (2016) | New algorithm to detect delay node, dropping node | Sink hole node detection | Detection rate, false positive rate | WSN |
| Xie et al. (2018) | Self-organization, dynamic topology | Data collection, security attacks | Accuracy, false tolerance | WSN |

detection strategies, uniform detection methods are proposed by Ma et al. (2019). Randomizing security polling with game-theoretical tools helps in countermeasures for the attacks. Interest flooding attacks for named data networks (NDN) are analyzed for reliability detection in Nguyen et al. (2019). Coupling NDN with internet protocol (IP) can sense false alarms and maintaining a trade-off between power and delay can help as countermeasures for flooding attacks. They can also be analyzed in terms of HTTP traffic, accuracy, and efficiency.

Intelligent sensor networks (ISN) and various attacks like side-channel attacks, power analysis attacks, leakage power analysis attacks are analyzed and countermeasures are addressed by Shanmugham et al. (2018). Performance metrics like asymptotic gains, coefficient deviations are considered for analysis in ISN. Information-centric networks (ICN) and the attacks in them are categorized into routing, naming, caching, and miscellaneous. Further, the attacks are related to the attributes, so confidentiality and privacy of the networks are addressed by maintaining security levels and solutions by AbdAllah et al. (2015). The attacks in WSN which are distributed DoS attacks, detection and solutions for those attacks are proposed by Abidoye et al. (2018). False messages are detected, compromised messages are identified and countermeasures are implemented. Coordinated attacks with bogus requests are sent and followed as one of the countermeasures. Various optimization techniques like ant colony optimization, moving average for replica detection algorithms are followed to detect node replicas in healthcare monitoring applications Anitha et al. (2021). Storage and control overheads are the performance metrics to analyze the algorithms with zero knowledge fingerprint-based replica detections.

Genetic algorithms and fruit fly algorithms are used for optimization in node capture attacks by Ruby Bhatt et al. (2020). Cost computations and efficiency are analyzed with these algorithms to use least resources and maximum node contributions in the network. Hilbert huang transform (HHT) is used by authors with spark-assisted correlation to detect the attacks Chen Hongsong et al. (2020). Spearman, Pearson coefficients are used as performance metrics to analyze low denial of service (DoS) in WSN. Power distribution in cyber-physical systems (CPS) to detect node attacks is proposed by Rong Fu et al. (2018). Firewalls and passwords are detected using graph theory and the network performance is analyzed using security, robustness, and throughput. Table 11.2 shows literature for attacks in low power and multi-disciplinary systems.

Worm hole attacks in smart metering applications are addressed using the merkle tree-based approach by Idris Khan et al. (2014). Simulations are carried out using NS2 and performance is checked using jitter, throughput, and delay of the network. Mechanisms for lossy networks and low power consumption are tried. A new architecture for IoT is designed to address the security attacks for IoT by Khanam et al. (2020). The countermeasures like power, memory for the processor are taken into consideration to handle the different types of attacks in IoT. Various attacks like line removal, line addition and line switching attacks of power systems are analyzed by Liang et al. (2017). NAA- Natural aggregation algorithm is used to find countermeasures to overcome these attacks and the performance of the algorithm is verified using Matlab in terms of energy balancing and efficiency. The security key algorithm is used to address the attacks in distribution automation systems by Lim et al. (2009). Cyber attacks are analyzed in terms of resource constraints and complex computations are reduced with real-time KEPCO test-bed and tests carried out. Decision making using multiple criteria, attacks graphs in a hierarchical approach. A multi-criteria decision-making for attacks in power control systems using a hybrid approach by Liu et al. (2010). Definition, construction, and

**TABLE 11.2**

Literature Survey for Attacks in Low Power and Multi-Disciplinary Systems

| Authors (year) | Algorithm/Technique | Merits and Demerits | Software and Metrics Used | Applications |
|---|---|---|---|---|
| Ma et al. (2019) | Randomizing security polling | Uniform detection, detection strategies | Game theoretical tools | Flooding attacks |
| Idris Khan et al. (2014) | Merkle tree-based algorithm | Mechanisms for low power and lossy network | Jitter, delay, and throughput using NS2 | Smart metering applications |
| Liang et al. (2017) | NAA (natural aggregation algorithm) | Line removal, addition and switching attacks | Energy, efficiency. Matlab is used | Power systems |
| Lim et al. (2010) | Security key algorithm for cyber attacks | Complex computations addressed | Resource constraints Korea Electric Power Corporation (KEPCO) test-bed | Distribution automation system |
| Liu et al. (2010) | Decision making based on multiple criteria, attacks | Hybrid approach for security degree | Definitions, Construction, vulnerability | Power control systems |
| Somani et al. (2017) | Scale inside-out approach | Reduction in attack downtime | Resources utilization factor | Cloud attacks |
| Varadharajan et al. (2018) | Architecture with policy for end-end services | Flow based and specific paths for transmissions | Context, routing information | SDN- software defined radio |
| Zheng et al. (2015) | Optimization algorithms for plain text attack detection | Highly efficient multi-carrier sub channel systems | Symbol error rate, mean square error | Physical layer security |

vulnerability are increased to increase the degree of security in the hybrid algorithm to propose countermeasures for power control system attacks. Security attacks in the cloud are addressed by Somani et al. (2017). Distributed (DDoS) denial of service is analyzed using an inside-out approach which helps in monitoring the resources utilization factor. The approach helps in reducing the attack downtime, servicing report time.

GNAVE algorithm is used to analyze the vulnerability in traffic to handle the attacks and increase security for wireless networks by Tague et al. (2008). Both routing analysis and joint security methods are followed, non-linear integer programming method is used to handle the attacks in wireless networks. An architecture with a policy for end-end services to handle attacks in software-defined radio networks (SDNs) by Varadharajan et al. (2018). Specific paths are predefined based on flow control and the countermeasures are analyzed in terms of routing information and context information. A new algorithm that detects delay node, message modification node and dropping node is proposed by Wazid et al. (2016) for hierarchical WSN. The nodes are divided into disjoint clusters and the sink attacker is detected by an active sensor meant for this purpose in each of these disjoint clusters. Table 11.3 shows a literature survey for attacks in IoT.

Optimization algorithms along with multi-carrier sub-channels systems are proposed to address plain text attacks in physical layer security by Zheng et al. (2014). Symbol error rate and least mean square error are the performance metrics considered for analyzing the behavior and detection of attacks. Various architectures are analyzed for data collection and addressing various attacks in WSN by Xie et al. (2018). Self-organization, dynamic topology architectures are analyzed for better performance in terms of accuracy, false tolerance.

**TABLE 11.3**

Literature Survey for Attacks in IoT

| Authors (year) | Algorithm/Technique | Merits and Demerits | Software and Metrics Used | Applications |
|---|---|---|---|---|
| Chander & Kumaravelan (2021) | ML & AI approaches | Detecting outliers/anomalies, abnormalities | Minimum power resources, computational capabilities | - |
| Aarika et al. (2020) | Classification phase of jamming, collision attacks | Perception layer security | Cryptography, error correction codes | IoT |
| Lounis et al. (2016) | Resource constraints for short range communications | Resource constrained defense mechanisms | Reducing anonymity, transmission power | Wireless technologies, IoT |
| Iqbal et al. (2020) | SDN protocol stack | Software defined security | Unifying all stakeholders platform | IoT security requirements |
| Nguyen et al. (2019) | Coupling NDN with IP | False alarm probability, trade off – power, delay | HTTP traffic, accuracy, efficiency | Named data networking (NDN) |
| Rekha Shanmugham et al. (2018) | LPA categorization and providing side channel security | Counter measures for medical sensors | Asymptotic gain, coefficient deviations | Intelligent sensor networks (ISN) |
| AbdAllah et al. (2015) | Categorization, relating attacks and attributes | Severity levels of attacks, security solutions | Privacy, confidentiality | Information centric networking (ICN) |
| Khanam et al. (2020) | New architecture for IoT | Counter measures for security attacks | Memory, power of the processor | IoT |
| Tague et al. (2008) | GNAVE algorithm | Non-linear integer programming | Network traffic | Wireless networks |
| Iqbal et al. (2020) | Elliptical curve cryptography - ECC | Functional, security requirements | Energy analysis using TinyOS | Node privacy in WSN |
| Nashwan (2021) | AAA (anonymous access authenticate) | High security, efficiency | Storage, computational cost analysis | WSN in big data environment |

## 11.3 FRAME Routing Algorithm

Flat routing helps in removing dead neighbor information from its own routing table rather than waiting for periodic network maintenance. Location-based routing does not involve complex calculations for forwarding packets. However, it should know its own location, one-hop neighbor's location and the location of the destination node. It deals purely with the location of nodes, but not on global topology information.

For routing between trusted parties within a network, minimum inter-domain intricacies, a flat rumor and multi-hop energy (FRAME) based routing with Bayesian network along with decision-making algorithm are proposed. The distributed algorithm and security research are the two other methods that concentrate on lack of interest, disobedience in inter-domain routing in a network among nodes.

### 11.3.1 FRAME Routing

Based on the flat routing methods and their characteristics, a novel flat rumor and multi-hop energy-based routing protocol is proposed. The proposed FRAME routing protocol forms a spanning tree in a hybrid-type Bayesian model assumed for any deployed network. The decision-making algorithm ensures the maximum weight in every step in order to select the best neighbor for a path to travel from the source node to the destination node. This in turn will create a pair-wise association between nodes, then form edges in the Bayesian tree.

FRAME routing involves the limited radius, free-radius routing in the network. Communication between the network server, radius server is managed by the user datagram protocol which is a transmission protocol based on the concept of free radius. Free radius protocol is a connectionless service whereas the radius limited routing is a connection-enabled service. This radius routing involves retransmissions and server availability. The network server acts as the radius client, which works as a proxy client. The server receives the user information, authenticates, and provides services.

### 11.3.2 Bayesian Network

Assuming Bayesian network model for any deployed nodes in fixed terrain is the novel construction when compared to state of art methodologies. It may be a hierarchical Bayesian network that keeps on varying based on a packet transmitted and its direction in network deployment.

In Figure 11.2, S is the source node; D is the destination node. L are leaf nodes, where the one-hop nodes are selected from the source using the proposed algorithm as represented by n = 1,2,3. The next-hop node is selected by verifying whether the next-hop node is not a leaf node – if it is a leaf node at all – and it should also be a destination node to be reached. Leaf nodes are the ones with no neighbor nodes within their communication range. Every next-hop node needs in its further steps or nodes in the path toward the destination.

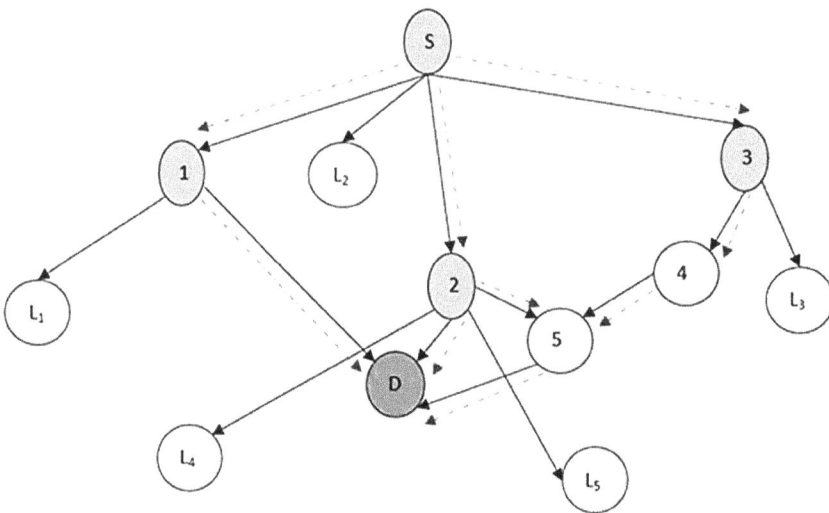

**FIGURE 11.2**
Hybrid Bayesian network.

The next hop node must be a destination node by itself if there are no further nodes are present. Hence, when a node is moving toward the destination in the shortest path, minimal energy consumption is assured.

Figure 11.3, shows the procedure to select next-hop node in the path from source to destination. If node A is the present node and node B & node C are the one-hop nodes from node A, then based on their node id and energy, the number of hops to the destination, the next-hop node is finalized in the path from source to destination.

Notations used in Figure 11.3

| | | |
|---|---|---|
| $NB_{ID}$ = Node B ID, | $NC_{ID}$ = Node C ID, | $NB_E$ = Node B Energy |
| $NC_E$ = Node C Energy, | $NB_H$ = Node B Hop count, | $NC_H$ = Node C Hop count |
| NXT_HOP = Next Hop | | |

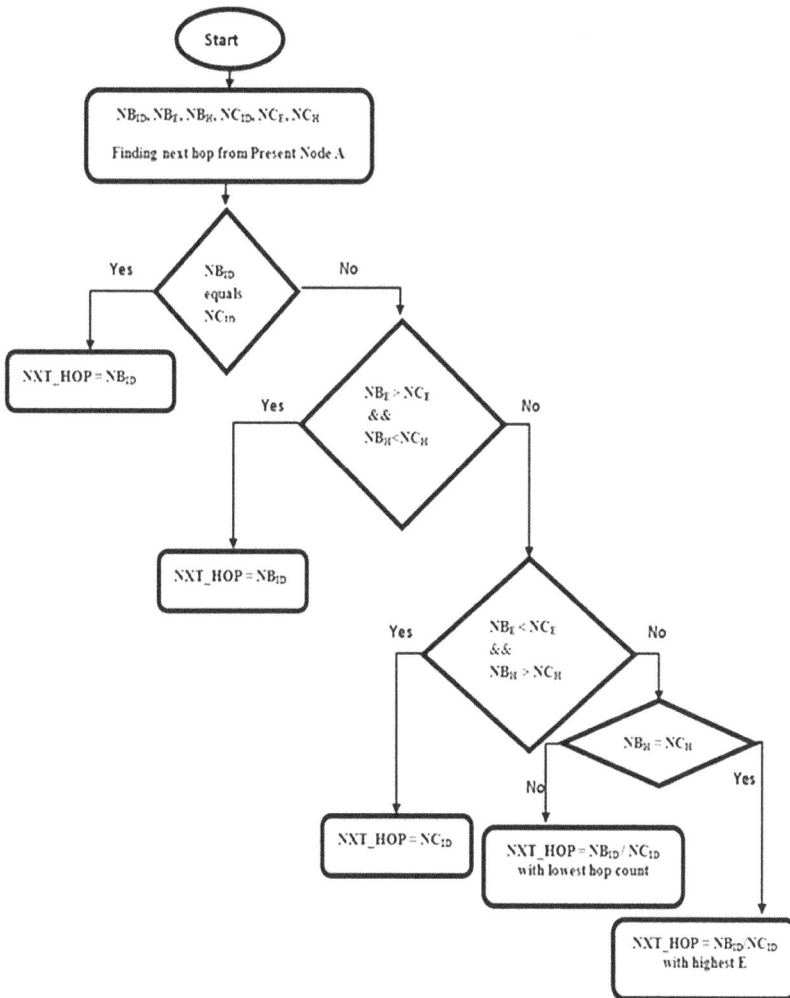

**FIGURE 11.3**
Flow chart of the decision-making (DM) algorithm.

If only one node is available in one hop from node A, then the condition is node B ID equals node C ID ($NB_{ID}$ equals $NC_{ID}$) which means only one node is available in one hop from node A and that will be the next hop toward destination from node A ($NXT\_HOP=NB_{ID}$). If two nodes node B and node C are available in one hop from node A, the energy and hop counts of node B and node C are compared. If node B has higher energy left and the minimum number of hops count toward destination than path via node C ($NB_E > NC_E$ && $NB_H < NC_H$), then node B will be finalized as next-hop from node A ($NXT\_HOP=NB_{ID}$).

The selection of the next hop is enabled for 10 hops in the network toward the destination while fixing every next-hop node. Hence, if node A can reach the destination with more energy and a minimum number of hops via node B, then node B will be finalized as the next hop. Instead, if node B has less left out energy and a greater number of hops to reach the destination than node C ($NB_E < NC_E$ && $NB_H > NC_H$), then node A will finalize node C as the next hop to reach the destination ($NXT\_HOP=NC_{ID}$). If the case is node B and node C has an equal number of hops to reach the destination, then node A can fix node B or node C which has higher energy left out ($NXT\_HOP=NB_{ID}/NC_{ID}$ with highest E) as a next-hop node to reach the destination. If the case is, node B and node C has an unequal number of hops to reach the destination, then node A can fix node B or node C which has a minimum number of hops ($NXT\_HOP=NB_{ID}/NC_{ID}$ with a minimum number of hops) as the next-hop node to reach the destination.

## 11.4 Results and Discussion

AODV protocol for ad hoc networks and FRAME routing for WSN are analyzed in this work. Simulations are carried out using NS-2.35. Various scenarios are simulated for performance analysis of AODV and FRAME routing protocols. The number of nodes is varied from 25 to 50 and then to 75. The terrain size is 500 × 500 and a droptail queue is used. Packet size is 512 bytes, omnidirectional antenna, and two-ray propagation models are used as shown in Table 11.4 simulation setup.

Routing protocols are analyzed with and without hackers. Hackers are the malicious nodes that interrupt the smooth flow of packets in the network and deteriorate the network performance.

### 11.4.1 AODV Without Attacks

Figure 11.4 shows throughput, jitter in AODV routing protocol. Throughput increases from 18223.4 Kbs to 21224.8 Kbps as the number of nodes increases from 25 to 50. Jitter decreases from 0.240616 ms to 0.150896 as the number of nodes increases from 25 to 50. Figure 11.5 shows delay and throughput when the speed of node increases. Delay changes from 0.319561 to 0.318795 ms as the speed increases from 20 to 40 m/s. Further delay changes to 0.00667465 ms as the speed increases to 60 m/s.

Figure 11.6 shows packets sent, energy consumption in AODV when speed of node increases. Packets sent increase from 37 to 41 as the speed of node increases from 40 to 60 m/s. Energy consumption decreases from 4.24163 to 1.7351 J as the speed of the node increases from 40 to 60 m/s. Figure 11.7 shows jitter, energy consumption in AODV routing when the speed of the node increases. Jitter decreases from 0.150896 to 0.148731 ms as

**TABLE 11.4**

Simulation Setup

| Parameter | Value |
|---|---|
| Number of nodes | 25, 50, 75 |
| Terrain | 500 × 500 |
| Queue type/queue size | DropTail/50 |
| Packet size | 512 bytes |
| Antenna type | Omni antenna |
| Propagation model | Two-ray ground |
| Routing protocol | AODV, FRAME* |
| MAC protocol | 802.11 |
| Speed (m/s) | 20, 40, 60 |
| Number of hackers | 1, 10, 20 |
| Initial energy | 20J/node |
| Simulation time | 10 s |

* Indicates proposed protocols

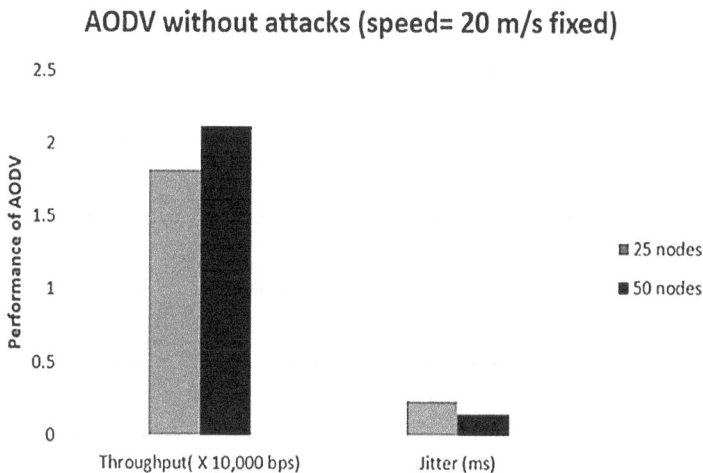

**FIGURE 11.4**
Throughput, Jitter in AODV routing.

the speed of nodes increase from 20 to 40 m/s. Energy consumption decreases from 4.39719 to 4.24163 J as the speed of nodes increases from 20 to 40 m/s.

## 11.4.2 AODV with Attacks

Figure 11.8 shows the deployment of 50 nodes with one hacker, Figure 11.9 shows the deployment of 50 nodes with 20 hackers. These are the attackers or malicious nodes, which interrupt the packet transmission and decrease the network performance.

Figure 11.10 shows packets sent, jitter in AODV routing with hackers. The packets sent increase 31–37 as the number of nodes increases from 25 to 50. Jitter decreases from 0.240616 to 0.150896 ms as the number of nodes increases from 25 to 50. Figure 11.11 shows control

## AODV without attacks (n=50 fixed)

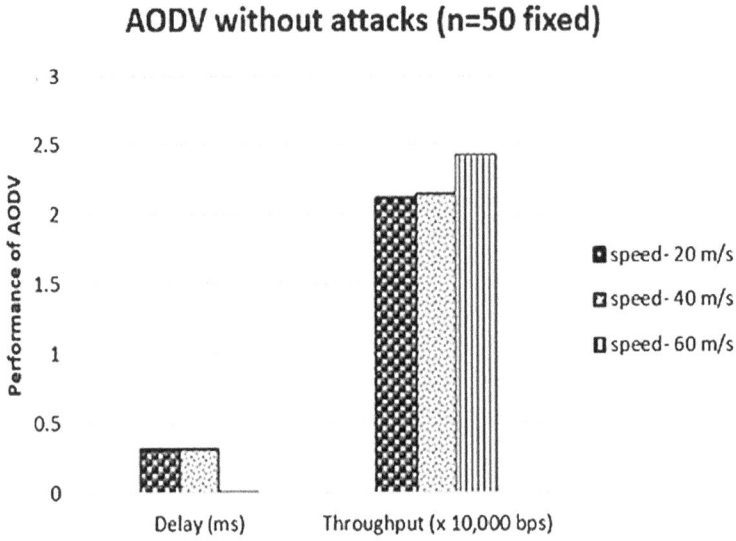

**FIGURE 11.5**
Delay, throughput in AODV routing.

## AODV without attacks (n=50)

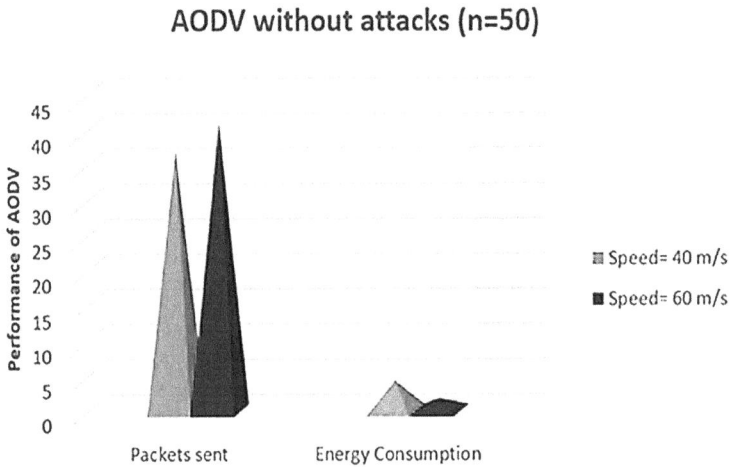

**FIGURE 11.6**
Packets sent, energy consumption in AODV routing.

overhead, throughput in AODV with attacks as the speed of the nodes increases. Control overhead decreases from 100 to 50 as the speed of nodes increase from 40 to 60 m/s. Throughput increases from 21458.6 to 24302.8 Kbps as the speed of nodes increase from 40 to 60 m/s. Figure 11.12. shows delay, Energy consumption in AODV with attacks as the speed of nodes increases. Delay decreases from 0.318795 to 0.00667465 ms as the speed of nodes increases from 40 to 60 m/s. Figure 11.13. show throughput, Energy consumption in AODV with attacks as the number of hackers increases. Throughput remains the same from 21224.8 to 21224.8 Kbps as the number of hackers increases from 1 to 10. Further throughput

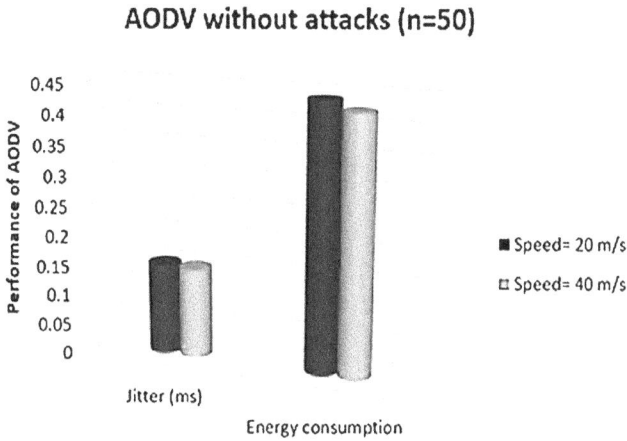

**FIGURE 11.7**
Jitter, energy consumption in AODV routing.

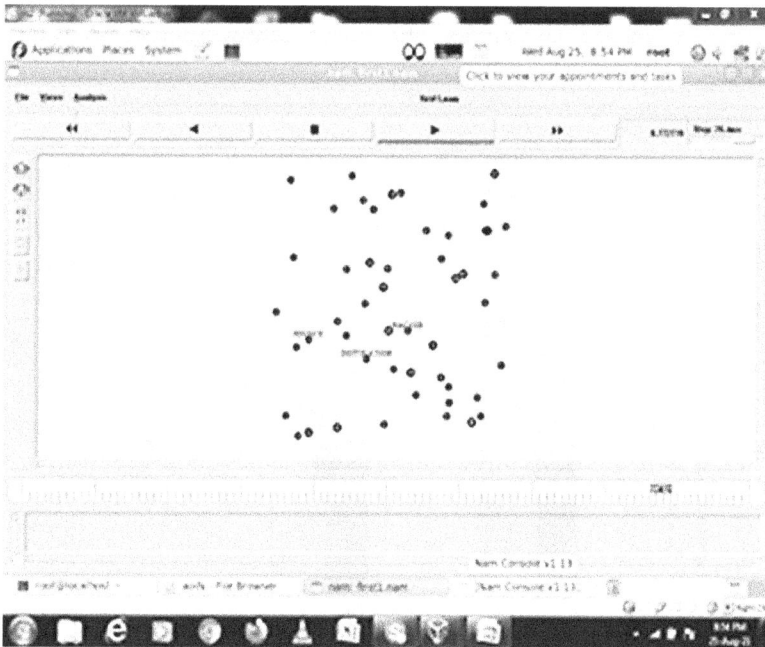

**FIGURE 11.8**
AODV routing with one hacker.

remains the same as 21224.8 Kbps as the number of hackers increases to 20. Energy consumption remains the same from 4.39719 to 4.39719 J as the number of hackers increases from 1 to 10. Further energy consumption remains the same as 4.39719 J as the number of hackers increases to 20. Hence, the AODV routing protocol is not efficient in handling hackers in the network.

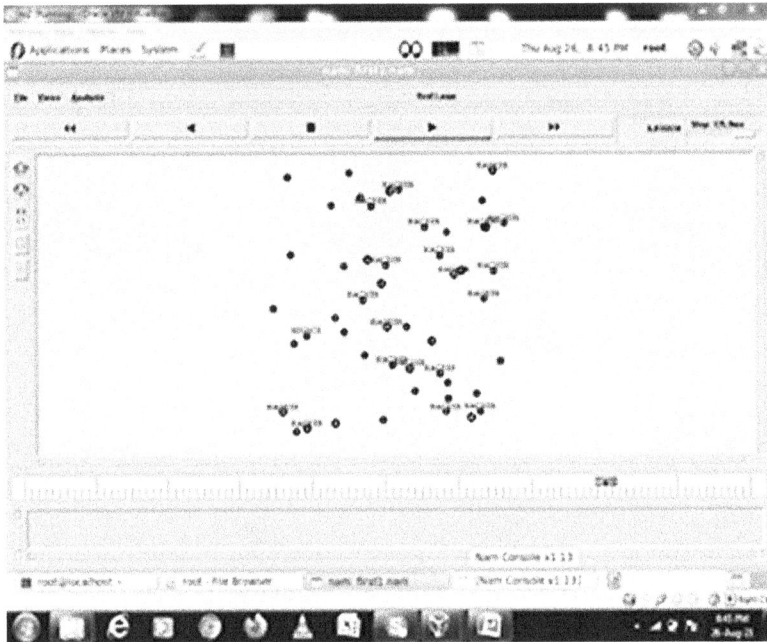

**FIGURE 11.9**
AODV routing with 20 hackers.

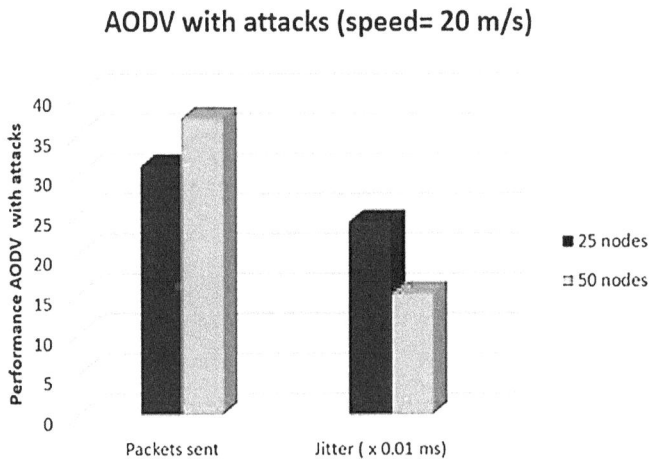

**FIGURE 11.10**
Packets sent, jitter in AODV with attacks.

### 11.4.3 FRAME Without Attacks

FRAME is the proposed routing protocol in this work which operates based on flat routing and particularly with the rumor routing protocol. Figure 11.14 shows the deployment of 25 nodes in 500 m² terrain and Figure 11.15 shows the deployment of 75 nodes in 500 m² terrain to operate with the FRAME routing protocol.

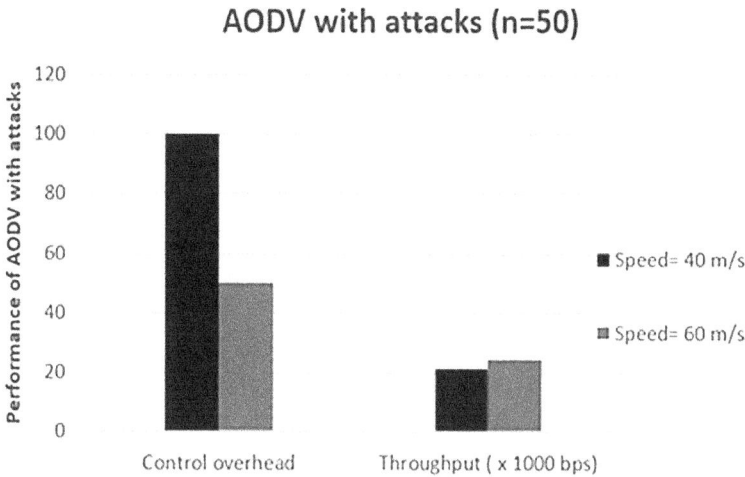

**FIGURE 11.11**
Control overhead, throughput in AODV with attacks.

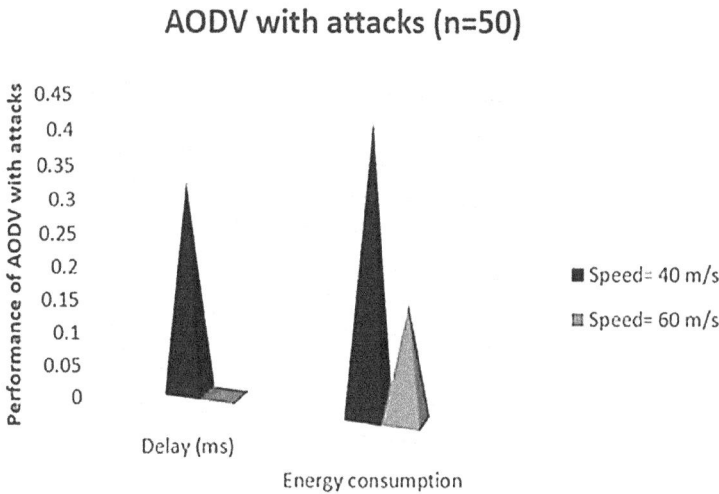

**FIGURE 11.12**
Delay, energy consumption in AODV with attacks.

Figure 11.16 shows PDR, throughput in FRAME routing with hackers/attacks as the number of nodes increases. PDR increases from 94.5055 to 98.9011 as the number of nodes increases from 25 to 75. Throughput increases from 16053.3 to 16800 Kbps as the number of nodes increases from 25 to 75. Figure 11.17 shows jitter, packets dropped in FRAME routing with hackers/attacks as the number of nodes increases. Jitter decreases from 0.102004 to 0.0995073 ms as the number of nodes increases from 25 to 75. Packets dropped from 5 to 1 as the number of nodes increase from 25 to 75.

Figure 11.18 shows packets dropped, PDR and delay in FRAME routing without attacks as the speed of nodes increases. Packets received from 84 to 87 as the speed of nodes increases from 20 to 60 m/s. PDR increases from 92.3077 to 95.6044 as the speed of nodes

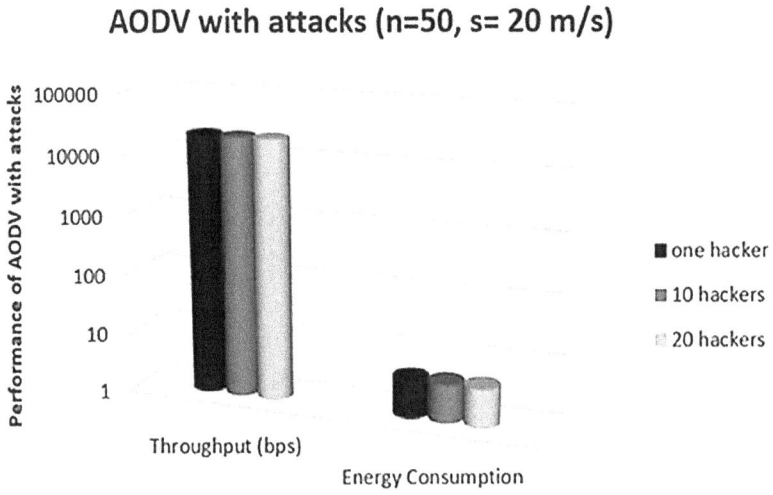

**FIGURE 11.13**
Throughput, energy consumption in AODV with attacks.

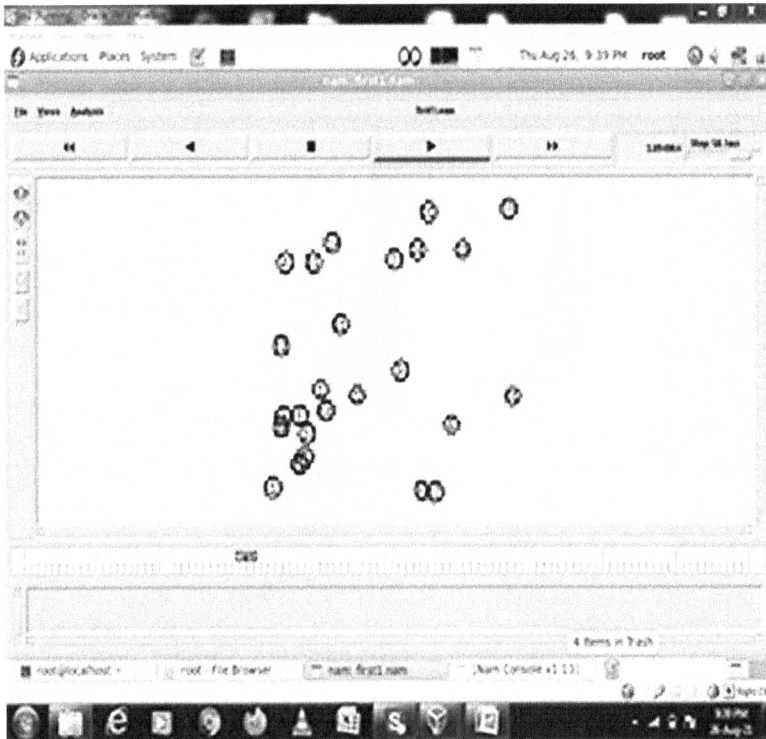

**FIGURE 11.14**
25 Nodes deployed in 500 m².

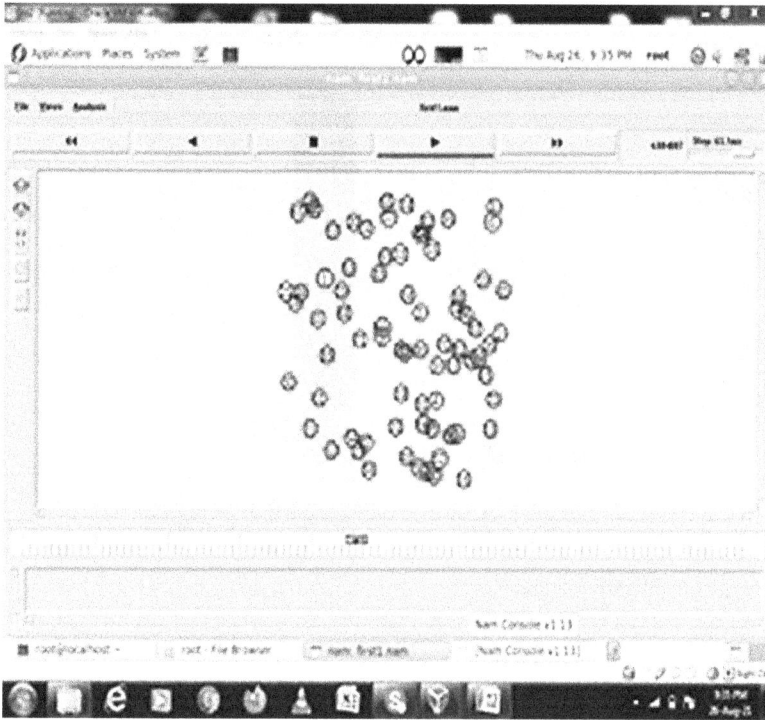

**FIGURE 11.15**
75 nodes deployed in 500 m².

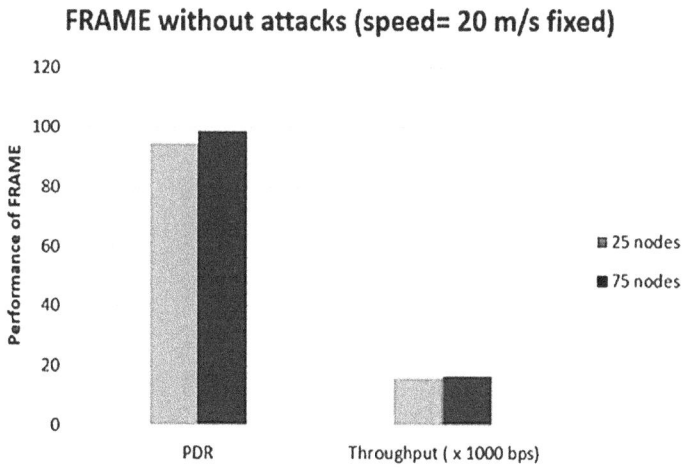

**FIGURE 11.16**
PDR and throughput in FRAME routing.

## FRAME without attacks (speed= 20 m/s fixed)

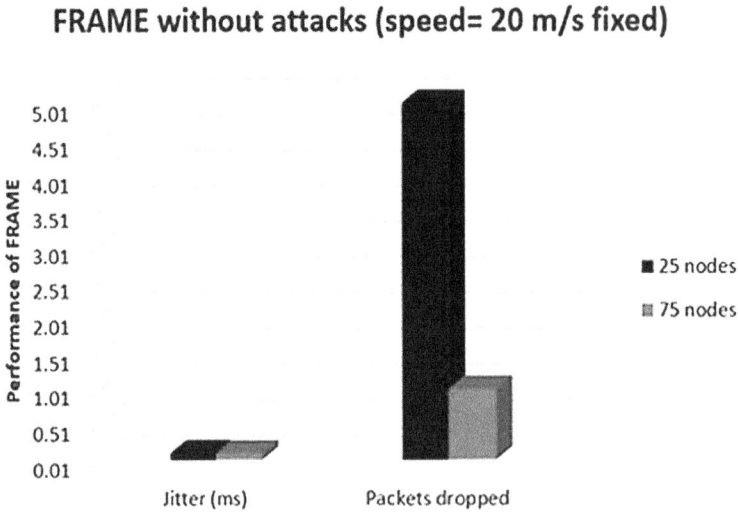

**FIGURE 11.17**
Jitter, packets dropped in FRAME routing.

## FRAME without attacks (n=50)

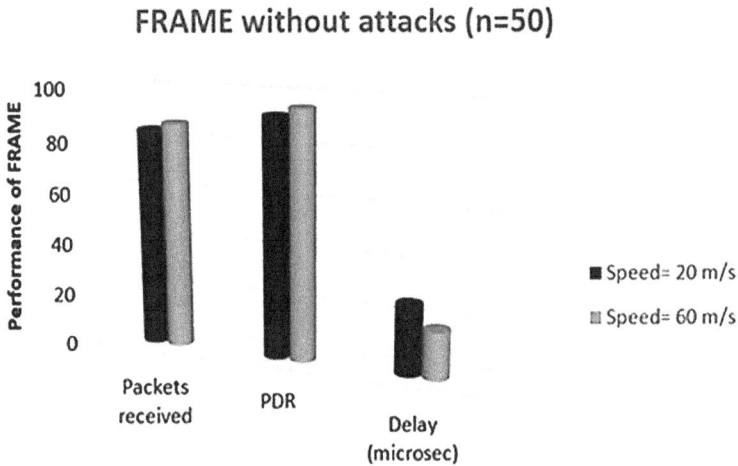

**FIGURE 11.18**
Packets dropped, PDR, delay in FRAME.

increases from 20 to 60 m/s. Delay decreases from 0.0269355 to 0.0184303 ms as the speed of nodes increases from 20 to 60 m/s. Figure 11.19 shows throughput, packets dropped, and energy consumption in FRAME without attacks when the speed of nodes increases. Throughput increases from 15680 to 16240 Kbps as the speed of nodes increases from 20 to 60 m/s. Packets dropped from 7 to 4 as the speed of nodes increased from 20 to 60 m/s. Energy consumption decreases from 8.86522 to 5.51951 J as the speed of nodes increases from 20 to 60 m/s.

## FRAME without attacks (n=50)

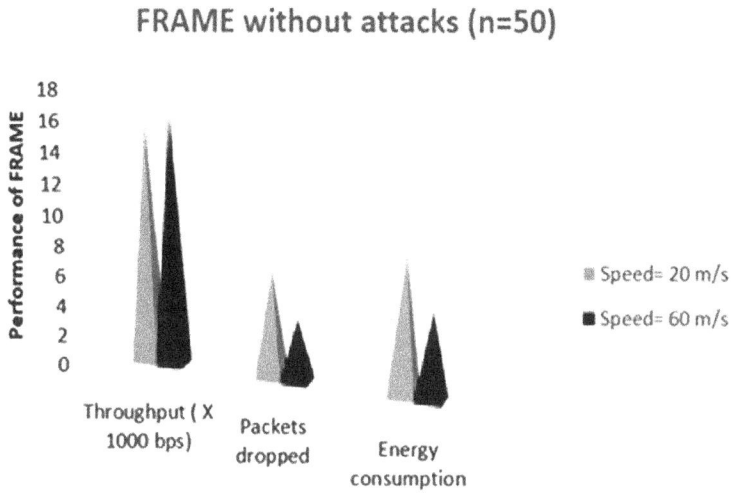

**FIGURE 11.19**
Throughput, packets dropped, energy consumption in FRAME routing

## FRAME with attacks
## (speed= 20 m/s, one hacker)

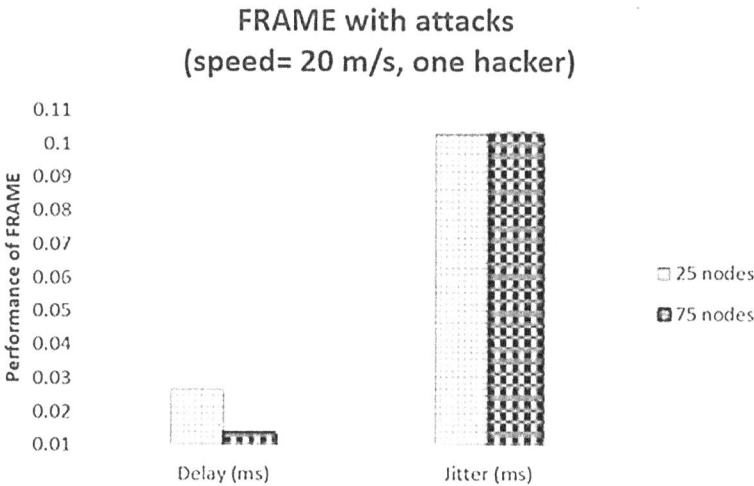

**FIGURE 11.20**
Delay, jitter in FRAME with attacks.

### 11.4.4  FRAME with Attacks

Figure 11.20 shows delay, jitter in FRAME routing with attacks as the number of nodes increases. Delay decreases from 0.0266591 to 0.0136859 ms as the number of nodes increases from 25 nodes to 75 nodes. Jitter decreases from 0.103147 to 0.102982 ms as the number of nodes increases from 25 nodes to 75 nodes. Figure 11.21 shows packets received, dropped, and PDR, energy consumption in FRAME with attacks as the speed of the nodes increases. Packets received increases from 87 to 89 as the speed of nodes increased from

## FRAME with attacks (n=50, one hacker)

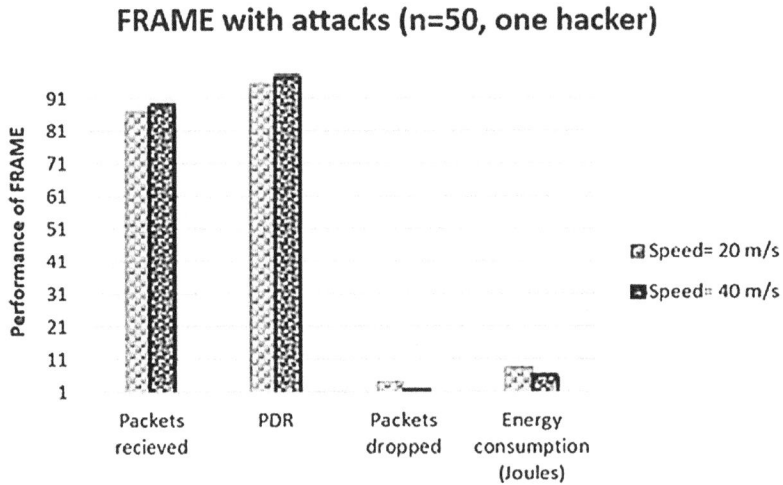

**FIGURE 11.21**
Packets received, dropped and PDR, energy consumption in FRAME with attacks.

## FRAME with attacks (n=50, speed= 20 m/s)

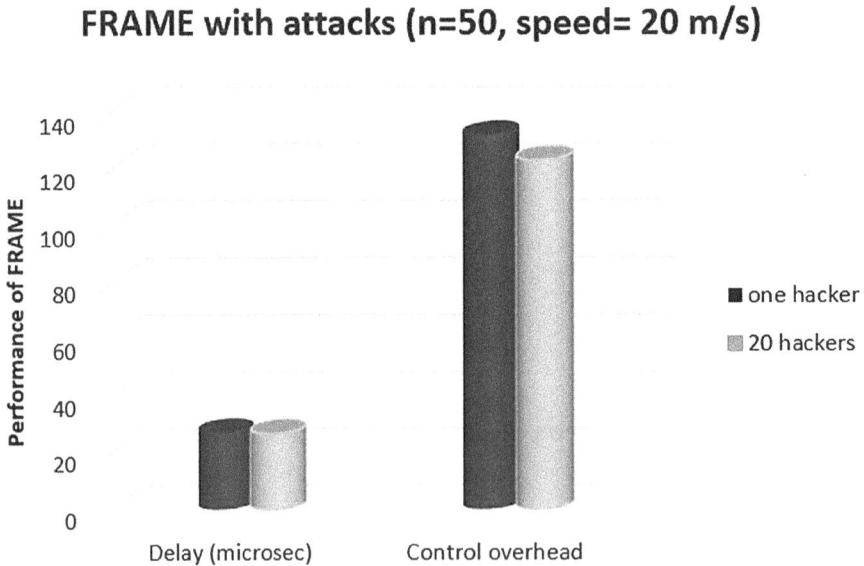

**FIGURE 11.22**
Delay, control overhead in FRAME with attacks.

20 to 40 m/s. Packets dropped from 4 to 2 as the speed of nodes increased from 20 to 40 m/s. PDR increases from 95.6044 to 97.8022 as the speed of nodes increases from 20 to 40 m/s. Energy consumption decreases from 8.53323 to 6.06368 J as the speed of nodes increases from 20 to 40 m/s.

Figure 11.22 shows delay, control overhead in FRAME with attacks as the number of hackers increases. Delay and control overhead remains the same 0.0268252 ms and 123,

## FRAME without/with attacks (n=50, speed= 40 m/s)

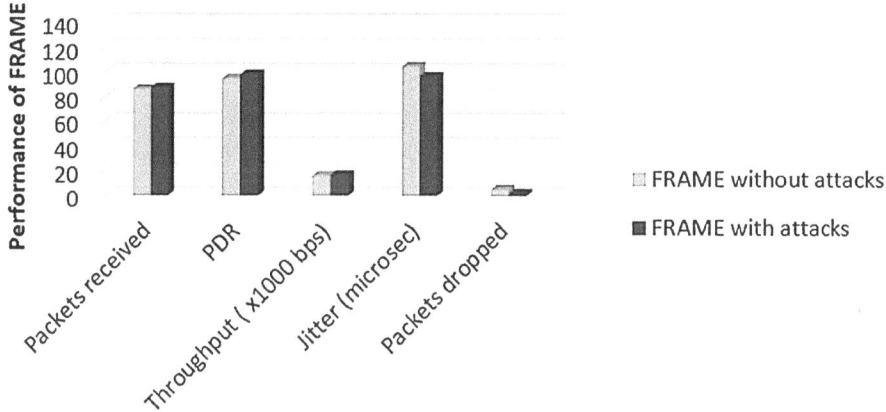

**FIGURE 11.23**
Packets received, dropped, and PDR, throughput, jitter in FRAME.

respectively. Figure 11.23 shows the performance of FRAME routing protocol with and without hackers. Packets received increases from 84 to 87 as the number of hackers increased from one to 20. PDR increases from 92.3077 to 95.6044 as the number of hackers increased from one to 20. Throughput increases from 15,680 to 16,240 Kbps as the number of hackers increased from one to 20. Jitter decreases from 0.106685 to 0.102961 ms as the number of hackers increased from one to 20. Packets dropped from 7 to 4 as the number of hackers increased from one to 20.

From all the simulations carried out, it is observed that FRAME routing performs better than AODV routing in terms of energy consumption, PDR, Jitter, delay control overhead, number of packets sent and received, and number of packets dropped.

## 11.5 Conclusion

### 11.5.1 Summary of the Work

- Conventional AODV routing protocols give better performance by increasing the speed of the nodes in the network. It is suitable for ad hoc network applications.
- Proposed FRAME routing consumes less energy, delay and has more throughput than conventional AODV. By increasing the number of nodes and speed of nodes in FRAME, improvement in throughput and PDR are obtained. Further increasing number of hackers/malicious nodes and an increasing number of nodes, speed of nodes helps in attaining PDR of 97.8%, control overhead decreases in addition to improvements in throughput.

**TABLE 11.5**

Performance of AODV Without and With Attacks

| Algorithm / Metrics | AODV With/Without Attacks (varying number of nodes; s = 20) | | AODV With/Without Attacks (varying speed s of nodes-; n = 50) | |
|---|---|---|---|---|
| | 25 nodes | 50 nodes | 20 m/s | 60 m/s |
| No of pkts sent send | 31 | 37 | 37 | 41* |
| No of pkts received | 31 | 37 | 37 | 41* |
| Pkt delivery ratio | 100 | 100 | 100 | 100 |
| Control overhead: | 25 | 100 | 100 | 50* |
| Normalized routing overheads | 0.806452 | 2.7027 | 2.7027 | 1.21951* |
| Delay | 0.00644481 | 0.319561 | 0.318795 | 0.0066746* |
| Throughput | 18223.4 | 21224.8* | 21458.6 | 24302.8* |
| Jitter | 0.240616 | 0.150896* | 0.148731 | 0.178585 |
| No. of pkts dropped | 0 | 0 | 0 | 0 |
| Total energy consumption | 1.0072 | 4.39719 | 4.24163 | 1.7351* |
| Avg. energy consumption | 0.0419667 | 0.0897386 | 0.086564 | 0.0354101* |
| Overall residual energy | 478.933 | 975.486 | 975.656 | 978.2 * |
| Avg. residual energy | 19.9555 | 19.9079 | 19.9113 | 19.9633* |

* Indicates better performance

## 11.5.2 Performance Analysis and Conclusion

Existing conventional AODV routing and FRAME routing are analyzed for better performance in this work. FRAME routing performs better than conventional AODV in terms of delay, control overhead. Energy consumption, PDR, packets received are shown in Table 5.1. In AODV routing protocol without attacks, by increasing the number of nodes and speed of nodes performance of the network can be improved in terms of throughput, jitter, delay, control overhead, and energy consumption. AODV protocol with attacks doesn't show better performance compared to AODV with attacks. Hence, it can be observed that AODV is not efficient in handling attacks or malicious nodes in the network.

Table 11.5 shows the performance of the AODV routing protocol with and without attacks. It can be observed that packets sent and packets received are increased from 37 to 41 by varying the speed of nodes from 20 to 60 m/s. Control overhead decreases from 100 to 50 and delay decreases from 0.318795 ms to 0.0066746 ms by varying speed of nodes from 20 to 60 m/s. Throughput increases from 21458.6 to 24302.8 Kbps, jitter decreases from 0.240616 to 0.150896 ms, and energy consumption reduces from 4.24163 to 1.7351 J by varying speed of nodes from 20 to 60 m/s in AODV routing protocol.

The performance of FRAME routing is shown in Table 11.6. The number of packets received increases from 87 to 89 when hackers are increased from one to 20.

Packets delivery ratio increases from 95.6044 to 97.8022 with one hacker in the network. Control overhead decreases from 190 to 123 in FRAME routing protocol with hackers. Delay changes from 0.0273812 to 0.026825 ms, throughput, jitter and number of packets dropped show better performance in FRAME routing with attacks. Hence, from all the simulations it is observed that FRAME routing with decision-making algorithm (DM) performance is better than AODV when there are attackers or malicious nodes in the network.

**TABLE 11.6**

Performance of FRAME without and with Attacks

| Algorithm / Metrics | FRAME without Attacks (s = 20) | FRAME without Attacks (speed s = 20) | FRAME with Attacks (s = 20, h = 1) | FRAME without Attacks (n = 50) | FRAME with Attacks (n = 50, h = 1) | FRAME with Attacks (n = 50, speed = 20) |
|---|---|---|---|---|---|---|
| | 25 nodes | 75 nodes | 75 nodes | 40 m/s | 40 m/s | hackers h = 20 |
| No of pkts send | 91 | 91 | 91 | 91 | 91 | 91 |
| No of pkts received | 86 | 90* | 87* | 86 | 89* | 87 |
| Pkt delivery ratio | 94.5055 | 98.9011* | 95.6044 | 94.5055 | 97.8022* | 95.6044 |
| Control overhead | 68 | 191 | 190* | 127 | 132 | 123* |
| Normalized routing overheads | 0.790698 | 2.12222 | 2.18391 | 1.47674 | 1.48315 | 1.41379* |
| Delay: | 0.0312048 | 0.013694* | 0.013685* | 0.013692 | 0.0273812 | 0.026825* |
| throughput | 16053.3 | 16800* | 16240* | 16053.3 | 16613.3* | 16240 |
| jitter | 0.102004 | 0.099507* | 0.102982 | 0.104239 | 0.096133* | 0.102961 |
| No of pkts dropped | 5 | 1* | 4 | 5 | 2* | 4 |
| Total energy consump: | 4.22572 | 7.94525 | 7.82616* | 5.04206 | 6.06368 | 8.53323 |
| Avg energy consumption: | 0.176072 | 0.107368 | 0.105759* | 0.102899 | 0.123749 | 0.174147 |
| Overall residual energy: | 235.774 | 732.055 | 732.157* | 484.958 | 483.921 | 481.448 |
| Avg res.energy: | 9.82393 | 9.89263 | 9.89401* | 9.8971 | 9.87593 | 9.82546 |

* Indicates better performance

### 11.5.3 Future Scope

- As an extension of this work, FRAME routing (network layer) can be checked with various MAC protocols (data link/MAC layer) for improving the performance of the network in with/without malicious nodes
- Other routing protocols for WSN can be designed for WSN

# References

Aarika, K., Bouhlal, M., Abdelouahid, R. A., Elfilali, S., & Benlahmar, E. (2020). Perception layer security in the internet of things. *Procedia Computer Science*, 175, 591–596.

AbdAllah, E. G., Hassanein, H. S., & Zulkernine, M. (2015). A survey of security attacks in information-centric networking. *IEEE Communications Surveys & Tutorials*, 17(3), 1441–1454.

Abidoye, A. P., & Obagbuwa, I. C. (2018). DDoS attacks in WSNs: detection and countermeasures. *IET Wireless Sensor Systems*, 8(2), 52–59.

Anitha, S., Jayanthi, P., & Chandrasekaran, V. (2021). An intelligent based healthcare security monitoring schemes for detection of node replication attack in wireless sensor networks. *Measurement*, 167, 108272.

Bhatt, R., Maheshwary, P., Shukla, P., Shukla, P., Shrivastava, M., & Changlani, S. (2020). Implementation of Fruit Fly Optimization Algorithm (FFOA) to escalate the attacking efficiency of node capture attack in Wireless Sensor Networks (WSN). *Computer Communications*, 149, 134–145.

Chander, B., & Kumaravelan, G. (2021). Outlier detection strategies for WSNs: A survey. *Journal of King Saud University-Computer and Information Sciences*. doi:10.1016/J.JKSUCI.2021.02.012.

Fotohi, R., Firoozi Bari, S., & Yusefi, M. (2020). Securing wireless sensor networks against denial-of-sleep attacks using RSA cryptography algorithm and interlock protocol. *International Journal of Communication Systems*, 33(4), e4234.

Fu, R., Huang, X., Xue, Y., Wu, Y., Tang, Y., & Yue, D. (2018). Security assessment for cyber physical distribution power system under intrusion attacks. *IEEE Access*, 7, 75615–75628.

Gao, Y., Ao, H., Feng, Z., Zhou, W., Hu, S., & Tang, W. (2018). Mobile network security and privacy in WSN. *Procedia Computer Science*, 129, 324–330.

Hongsong, C., Caixia, M., Zhongchuan, F., & Lee, C. H. (2020). Novel LDoS attack detection by Spark-assisted correlation analysis approach in wireless sensor network. *IET Information Security*, 14(4), 452–458.

Huanan, Z., Suping, X., & Jiannan, W. (2021). Security and application of wireless sensor network. *Procedia Computer Science*, 183, 486–492.

Idris Khan, F., Shon, T., Lee, T., & Kim, K. H. (2014). Merkle tree-based wormhole attack avoidance mechanism in low power and lossy network based networks. *Security and Communication Networks*, 7(8), 1292–1309.

Iqbal, U., & Mir, A. H. (2020). Secure and practical access control mechanism for WSN with node privacy. *Journal of King Saud University-Computer and Information Sciences*. doi: 10.1016/j.jksuci.2020.05.010.

Iqbal, W., Abbas, H., Daneshmand, M., Rauf, B., & Bangash, Y. A. (2020). An in-depth analysis of IoT security requirements, challenges, and their countermeasures via software-defined security. *IEEE Internet of Things Journal*, 7(10), 10250–10276.

Keerthika, M., & Shanmugapriya, D. (2021). Wireless sensor networks: active and passive attacks vulnerabilities and countermeasures. *Global Transitions Proceedings*, 2(2), 362–367.

Khanam, S., Ahmedy, I. B., Idris, M. Y. I., Jaward, M. H., & Sabri, A. Q. B. M. (2020). A survey of security challenges, attacks taxonomy and advanced countermeasures in the internet of things. *IEEE Access*, 8, 219709–219743.

Liang, G., Weller, S. R., Zhao, J., Luo, F., & Dong, Z. Y. (2017). A framework for cyber-topology attacks: Line-switching and new attack scenarios. *IEEE Transactions on Smart Grid*, 10(2), 1704–1712.

Lim, I. H., Hong, S., Choi, M. S., Lee, S. J., Kim, T. W., Lee, S. W., & Ha, B. N. (2009). Security protocols against cyber attacks in the distribution automation system. *IEEE Transactions on Power Delivery*, 25(1), 448–455.

Liu, N., Zhang, J., Zhang, H., & Liu, W. (2010). Security assessment for communication networks of power control systems using attack graph and MCDM. *IEEE Transactions on Power Delivery*, 25(3), 1492–1500.

Lounis, K., & Zulkernine, M. (2020). Attacks and defenses in short-range wireless technologies for IoT. *IEEE Access*, 8, 88892–88932.

Ma, X., An, B., Zhao, M., Luo, X., Xue, L., Li, Z.,… & Guan, X. (2019). Randomized security patrolling for link flooding attack detection. *IEEE Transactions on Dependable and Secure Computing*, 17(4), 795–812.

Nashwan, S. (2021). AAA-WSN: Anonymous access authentication scheme for wireless sensor networks in big data environment. *Egyptian Informatics Journal*, 22(1), 15–26.

Nguyen, T., Mai, H. L., Cogranne, R., Doyen, G., Mallouli, W., Nguyen, L.,… & Festor, O. (2019). Reliable detection of interest flooding attack in real deployment of named data networking. *IEEE Transactions on Information Forensics and Security*, 14(9), 2470–2485.

Shanmugham, S. R., & Paramasivam, S. (2018). Survey on power analysis attacks and its impact on intelligent sensor networks. *IET Wireless Sensor Systems*, 8(6), 295–304.

Somani, G., Gaur, M. S., Sanghi, D., Conti, M., & Rajarajan, M. (2017). Scale inside-out: Rapid mitigation of cloud DDoS attacks. *IEEE Transactions on Dependable and Secure Computing*, 15(6), 959–973.

Tague, P., Slater, D., Rogers, J., & Poovendran, R. (2008). Evaluating the vulnerability of network traffic using joint security and routing analysis. *IEEE Transactions on Dependable and Secure Computing*, 6(2), 111–123.

Varadharajan, V., Karmakar, K., Tupakula, U., & Hitchens, M. (2018). A policy-based security architecture for software-defined networks. *IEEE Transactions on Information Forensics and Security*, 14(4), 897–912.

Wazid, M., Das, A. K., Kumari, S., & Khan, M. K. (2016). Design of sinkhole node detection mechanism for hierarchical wireless sensor networks. *Security and Communication Networks*, 9(17), 4596–4614.

Xie, H., Yan, Z., Yao, Z., & Atiquzzaman, M. (2018). Data collection for security measurement in wireless sensor networks: A survey. *IEEE Internet of Things Journal*, 6(2), 2205–2224.

Zheng, Y., Schulz, M., Lou, W., Hou, Y. T., & Hollick, M. (2014). Highly efficient known-plaintext attacks against orthogonal blinding based physical layer security. *IEEE Wireless Communications Letters*, 4(1), 34–37.

# 12

---

*A Blockchain Security Management Based on Rehashing Shift Code Rail Encryption Using Circular Shift Round Random Padding Key for Decentralized Cloud Environment*

**K. Ganga Devi and R. Renuga Devi**

*Vels Institute of Science, Technology and Advanced Studies (VISTAS), Chennai, India*

## CONTENTS

---

## 12.1 Introduction

As well as sharing Blockchain technology is a database shared between P2P peer-to-peer network deals the security problems. It is the sequence of modules that are connected over time, ensuring the public-key encryption is stable and holds intuitive transactions sealed by the network community. Blockchain on non-interactive data logs managed by a cluster of references to computers, in other words, a series (timestamp). Each of these data

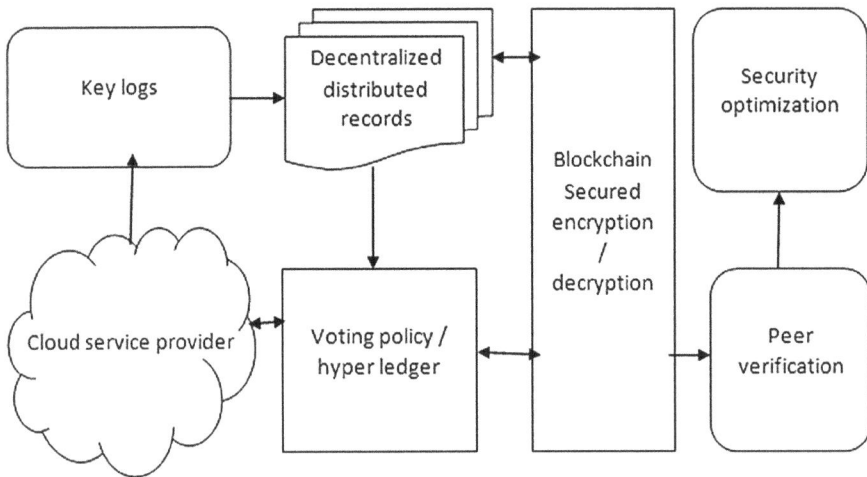

**FIGURE 12.1**
Process of decentralized vote aggregated transaction in the Blockchain.

blocks has a standard as it is surrounded by encryption policies. Besides, Blockchain can be divided into two types: private and public Blockchain. Private Blockchain contributes to a network that is restricted in terms of access right where it is most important to reduce the number of users entering.

Blockchain possibilities still need to be explored, and its security model will be improved, especially in the area of storage security. Different platforms have not yet been hit by extreme cloud conditions based on data integrity. In this paper, researchers who use the platform in extreme situations and use Blockchain technology to propose more sustainable and secure system solutions investigate the availability of data on distributed currencies. Figure 12.1 shows the process of decentralized vote aggregated transactions in the Blockchain.

The Blockchain cloud is thin compared to traditional cloud computing infrastructure. It was implemented to enable the Bitcoin cryptocurrency payment system. Blockchain enables incredible parties to exchange, process, and create permanent transparent records without relying on a power center.

Not enough to prevent hash functions being damaged. Can effectively count hundreds of hashes to thousands of calculations scams a volume and a block that takes a few seconds to recalculate all the hashes needed for the home. Therefore, to protect more than stamp blockchains, use (transactional) proof of work. Captive is a mechanism that minimizes the appearance of new blocks. If you change the modules, this mechanism makes the modules difficult to work with, as they all need to be recalculated for the following modules.

It is provided by Blockchain that maintains similar functionality and store transactions but without the need for a third party. As a transaction with Pareto decentralization, each participating user holds a copy of the original Pareto in Blockchain Pareota, where the power center verified if the problem is solved. Any participating user can request additional transactions. However, the user who has confirmed the transaction will be included in the majority of the Blockchain network. Automatic testing by security and fast payrolls will help to significantly interrupt the transaction and transaction and make it reliable for every user.

## 12.2 Related Work

Blockchain is a technology that is supposed to be the best cloud computing. Blockchain will overcome security issues in the computing cloud [1,2]. The purpose of this study is to study and compare the various issues of the cloud environment and security issues using Blockchain.

Blockchain technology is considered to be an ideal fit for enhancing existing computer systems in a variety of ways [3]. Cloud computing is one of the most widely used network compatible technologies and is used only professionally by many cloud service models. It has great potential to improve both functionality/performance and Blockchain technology to improve security/privacy by melting existing cloud systems [4]. Blockchain Technology Inserts The cloud data center is currently widely questioned as to how to allow a redesign of cloud solutions. These are the latest initiatives of the trial addresses and trial Blockchain and cloud technology merger.

The three techniques cover this piece approximately in dimensions. First, we are reviewing the growing cloud-related Blockchain service model concerned with the service model, Blockchain-Powered Service: Service (Path); Second, security is considered an important technical dimension of this job, evaluating both access control and searchable encryption methods [5]. The Blockchain support/participants and cloud data center performance from a hardware and software perspective. Key effects of this study Blockchain-enabled cloud data center reconstructions will provide theoretical support for future pointers.

Recently, Blockchain technology [6,7] has shown its legacy of resolving reliability issues. Among these programs is Blockchain, a shared account book, data center maintained by a trustless peer-to-peer network. The code stored in the Blockchain smart contract [8] is used to implement the features of unstructured data, data storage, and data recovery. If the results are accurate, all movements are made so that through all the nodes in their network you can be assured that the nodes are very true honest.

However, such extensions suffer from privacy and capacity issues. In particular, from the interim results, that is, each key relevant data service record [9–11] is highlighted. The disability of such data should not raise a privacy complaint. Also, one of the service friends has to execute a single search request. Due to that, the predictable cost of processing multiple times of such words can be quite burdensome even for some keywords [12–14], most or all of the data can be displayed. Interim results lead to significant financial costs for the vast majority of the public as smart contracts are written by their service equivalents [15]. After all the keyword results have been calculated, the additional calculation of service counterparts will calculate the cut-off of the results requested.

The Blockchain is maintained and distributed by distributors who have distributed the entirety of previous transactions. It is provided by Blockchain that maintains similar functionality and store transactions [16] but without the need for a third party. As a transaction with ledger decentralization, each participating user holds a copy of the original ledger in Blockchain Pareota where the power center verifies if the problem is solved.

Any participating user can request additional transactions. However, the user who has confirmed the transaction will be included in the majority of the Blockchain network [17]. Automated testing modules by security and fast payrolls will help to significantly interrupt the transaction and make it credible to every user.

The latest developments in computing quantum, most of the widely used encryptions are serious about classical encryption, as some problems are based on the hardness that can be solved efficiently using quantum computers. It's the cause. Therefore, the study of post-quantum cryptography [18] has taken a great leap forward. Destruction would be enormous, affecting the security of public-key encryption on quantum computers.

Ellipse curve encryption (ECC), a general key encryption approach, is often used for Blockchain applications. Quantum computers break keys easily, thus forging the curve, using the Beach algorithm [19] variant as a security boost that could block Blockchain based on the corrupt digital signature of Blockchain and SO-ECC for each of those transactions.

## 12.3 Problem Consideration

Security, on the other hand, has long been an important issue for customers, especially in the banking and government sectors. The biggest challenge for any comprehensive access control solution for outsourced data is the ability to handle resource requests following species conservation policies to protect user-information security [20]. Many solutions have been proposed but most of them do not consider user-owner policies and user access methods to protect privacy as an important aspect of users need right authentication.

When you purchase and upgrade (PP) based on the evidence of a consensus algorithm whether it is working by the general public or by publications. Anyone can read, write, and send transactions. The buyer then confirms the transaction via an encryption field that will deliver or create the transaction or block. Transactions on a shared database are Blockchain guaranteed and delivered to workers. Vending the seller via trustless will also receive transactions. Use stock work proof and other consensus mechanisms proof to protect these handbooks.

## 12.4 Materials and Methods for Improved Blockchain Security

Control access data storage policies, processes, and control functions contain the data service offerings required for each distribution. Total control to maintain an environment of data storage. Selective security, that supports specific controls and control structures that are capable of storing data file characterization and availability, users, processes, and technology required for content. To propose a rehashing shift code rail encryption (RSCRE) using circular shift round random padding key for decentralized cloud environment (CSR2PK). The figure given below shows the Blockchain security [21–24].

Control access data storage policies, processes, and control functions contain the data service offerings required for each distribution, e.g., total control to maintain an environment of data storage. Figure 12.2 shows the architecture for the Blockchain security RSCRE-CSR2PK working principle. Selective security that supports specific controls and control structures that are capable of storing data file characterization and availability, users, processes, and technology required for content.

The digital identity blockchain (DIB) gives the identity owner more control over personal information while eliminating digital identity enterprise concerns. It provides a platform for running and verification of integrated identities by providers on the same platform. This strength forms a decentralized system that is activated by blockchains as information point transfer nodes. Identifiers, providers, and auditors to maintain privacy by adhering to the code of conduct. The registry is a database that explicitly permits items.

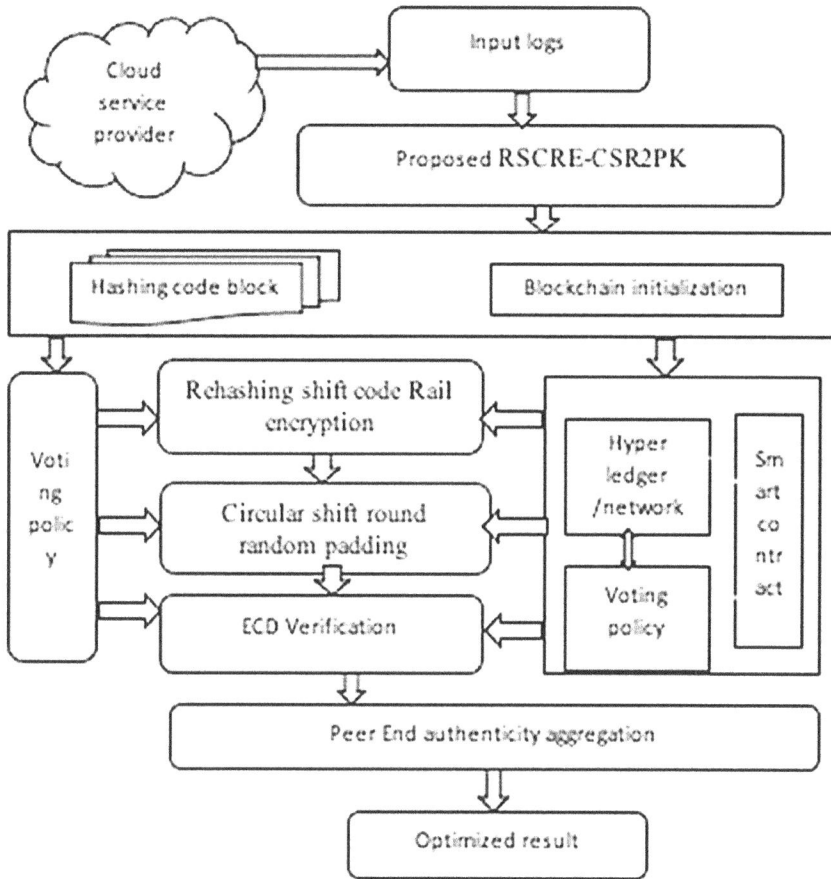

**FIGURE 12.2**
Architecture for Blockchain security RSCRE-CSR2PK.

Many public records are consolidated and maintained by public authorities to ensure the authenticity of any items registered with the authority.

## 12.5 Hyperledger Transaction Security

Hyperledger fabric (HLF) creates security and is an open-source implementation of a distributed parade platform that drives smart contracts. This article presents a hyperledger performance model. We set up the HLF network in our lab and run it using workload hyperledger calipers. From the results of our study, we find that most (and) policies affect the number of peers to complete the approval process at that time. Deployment services and writing performance issues on the parchment can also be limited by using larger volume sizes, even though increased inactivity is the norm. Involvement colleagues, transaction validation testing (using the validation system Blockchain code (VSCC)) is a time-consuming process that can be easily parallelized to minimize its performance impact. Its performance needs to absorb the impact of the block arrival explosion.

## 12.6 Smart Contracting Security

The first challenge is to set a piece of hidden information for access contract-based limit terms in smart contracts dealing. That said, service counterparts' smart deals contribute their predictive resources to manage the status of devices. With each operation of a smart contract, for example, both numbers on the partition disk add savings and, due to which a certain amount of money is spent such work is not free. To ensure the validity of smart contracts, tariff control rules that cost a single transaction are minimal. During the setup phase, the data encryption service is assigned to several service peers.

In particular, our algorithm creates a random code key to hashing the data order in linear, reversing, and encrypting key identifier pairs and tags supported multiple times. To prevent the processing of the system obtained for violating the rules of customs controls, we can create with each identifier key pair and use it to analyze the indexing point of key creation order for data encryption can be controlled by a reversing decryption unit. Create a way shifting standard hashing policy of each round the data contain the number of bytes be relocated by encoded format. After that, cutting the encryption information depends on the block by code of retaining function with verified key updated into stored cipher file. Finally, the encryption key identifier pair and identify are roughly mixed to prevent data leakage.

Input: Essential Modulus Bit Length, $L$.
Output: An random key pair ($(M, f)$, e) where N is the modulus, the creation of two primes ($M=sr$) not beyond l bits in length; f is the public advocate, e number less than and co-prime to $(r-1)(s-1)$; and d is the private exponent such that $ed \equiv 1 \bmod (r-1)(s-1)$.
Select a value of e from exponential terms

Repeat
$$R \leftarrow nonprime\ (L/2)$$
Until $(Rmode) \neq 1$

Repeat
$$S \leftarrow nonprime\ (1 - L/2)$$
Until $(s\ mode) \neq 1$
$$M \leftarrow sr$$
$$L \leftarrow (r\text{-}1)(s\text{-}1)$$
$$D \leftarrow mod\ inv\ (f, L)$$

Return $(M, f, e)$

Where, Advanced Encryption RSA (AERSA), R, S- encrypting variable-key length, data transmission detection technology can assist in the classification of traffic to establish the first data encryption at present. Then, by creating a second file with another type of data file from the first file, the algorithm implements the controls already implemented using AERSA. The result is that there are at least two files common paths between each data parallel to the network security.

The second challenge lies in designing multiple search methods and financial mechanisms to attain original data security. As mentioned earlier, the causal approach is always ineffective in terms of delays and financial costs as temporary consequences. In this process, a multi-aggregated voting key policy is generated based on our insights that make these terms look very informative. First, create a key identifier tag for each identifier at the system level.

On the other hand, we secure key exchange peer end verification key that is encoded to store in the database. During the multiple exploration stage, search for intermittent words for search queries, filter information, and use a bloom filter to use it. Keep in mind that the words you choose are intermittent because most of the words stay away from the results. Finally, we can use tags with key pairs to ensure that each identifier search request meets each candidate's identifier.

## 12.7 Blockchain Initialization

It manages an Amazon web service created with the help of hyper ledgers and network modules to protect Blockchain data. The Blockchain can be initiated via the following steps:

Initialize Setup. {(idi, wi) | i = 1, 2, $\cdots$, n}, = is a list of key pairs of data identifier service counterparts L identifier, information is outsourced by W. Each Wi is the string at the elongated length $\in$ {0, 1} * is a string of constant length $\mu(0, 1)$. The Clair text pointed the shuffle pair length, weight be formulated as L, M, and N which record the identifier key pairs, service id, service request, cloud request correspondingly.

In assigned data aggregated by the data owner, Info W is a package identifier for the key pair {ID, Wa as the core of the keywords added to its keywords.

Data Owner Keywords are a collection of what information W removes from the identifier key pairs {ID, WD}. Search. Data Owner W $\in$ W Existing WS = {Link W1, W2, $\bullet\bullet\bullet$, Week} Service Identifiers All Identifiers ID $\in$Ws must find and send (ID, VAT) as formed at the peer end, enable access requests at service partner transactions, and close transactions within Blockchain. If the transaction is then confirmed by Blockchain, the data aggregate form voting policy begins the smart contract, which makes an action to obtain a smart contract of service mates. Finally, the results are aggregated with data owners, depending on the status of their smart contracts.

## 12.8 Padding Key Substitution-Algorithm

An important example of an advanced monoalphabetic replacement is the alphabetical order variable,

Step 1: Encryption substitution:
   i)   ABCDEFGHIJKLMNOPQRSTUVWXYZ
   ii)  UNIVERSTABCDFGHJKLMOPQWXYZ
Step 2: Find each letter of nature in the first row of this table of Encryption, and then replace it with the letter below it. In our example, the size is:
   iii) ENGLI SHAST RONOM ERWIL LIAML ASSEL LDISC over DTRIT
   iv)  EGSDA MTUMO LHGHF ELWAD DAUFD UMMED DVAMI HQELE VOLAO HG

Step 3:  For decoding, we use the reverse order variable, given by the table
   v)   ABCDEFGHIJKLMNOPQRSTUVWXYZ
   vi)  IJKLEMNOCPQRSBTUVFGHADWXYZ
Step 4:  The central permutation must be a converter: the exchange is two rows and the first row of sorts.

Change Password contains a simple change (or change of characters) of the central location. All transliterations use advanced single-character passwords. So, we have my alphabet! In practice, it is a key that usually uses a limited selection of alternatives. With a "key" all letters found in the previous stages are discarded, and all unused messages in alphabetical order are merged.

## 12.9  Hash Key Generation on Padding

Role key Holomorphic Encryption Algorithm

Step 1:
Encryption and Decryption (no of keys in Generated)
Start
Number and alphabetic formal [0-9, a-z], Formal = in
Step 2:
 If(id == 0)

 {
Add Round Key (formal, w [0, Key-1])
For pattern key = 1 step 1
Split Number and alphabets (formal)
Formalized to shift the data rows
Shift to randomize mixing row and columns
Add pattern Key (check if any key
Compute formalized terms, W [forum pattern Securitykey* (Reqpattern Key +1)-1])
End for
Add pattern Key format (formalization, [w *NA, with doc(NA+1)*NA-1])
}
Else
{
returnOut = formalized data forum
}
Step 3: Return Formalized key
End

The extensive function of the encrypting algorithm varies depending on the variable and the key.
   Without the key, the document cannot be improved or decrypt.

## 12.10 Circular Shifting Block Generation

The data owner must send the database of encryption service colleagues. However, the encryption information must be truncated before it can be sent due to the cost of the rules governing smart contracts. Generally, the vote aggregator performed smart contracts by is sent at a specific cost per activity in a smart contract.

Right-left state matrix columns with rotation index elements can point to a specific rotation. State team bits do not change to the starting state of the sequence. Following the second row principle, finally rotate the number of consecutive bytes 1 each. Block size team along with 256 bits in different ways basically matrix hashing alignment.

The module, which has a size based on the array state, is the same as the conversion method, resulting in a byte change of 256 bits, respectively. As a result, the number of data that can be linked to each transaction is limited. In our protocol, each identifier key pair, i.e., non-generative data, is found to be designed for key identifier pairs and is subject to markers. Keyword tag pair and size of Clair text, for example, protect 512-bit and 256-bit, respectively.

---

**ALGORITHM**

**Input:** column matrix, row matrix (CR) Matrix encoder, shifter cipher row(Scr)
**Output:** Encoded index Ek.

Step 1: Initialize the Ciphertext on matrix
Step 2: compute index for each session creta index Si
         Matrix encoder index Mei = shifter column point → Scr.
         Compute for all index iteration Mei → Scr
             Circular Shift MI->CR to the user.
             Random shuffle key index shift point Scr(Rsp)
         Process End.
Return shift order RfS
         End
Step 3: While Mei → matrix transfoirmation0
         Index → (Scr) + current key;
         End
Step 4: Compute the new session order to a circular shift
         Go to step2 to repeat the end.
Step 5: End process

---

## 12.11 Rehashing Shift Code Rail Encryption

Encryption plans based on multiple receiving unity can be found in the Identity-Based Broadcasting Encryption (IBBE) program rather than one receiving as a generalization. With broadcast encryption, users can identify their identity by revealing their public key.

In many receiver systems, IBBE has proven to be a powerful way to provide data security and privacy. In this mode, the sender offers the package called user offer package, encrypted message broadcast. There can be many offer boxes with different numbers.

---

Input: preprocessed set Ps, service level trusted encryption SEt., content shift block Chs,

Output: encrypted data with public-key set PKs.

Start

Step 1:   input the record to read content to shift encryption

For each record type Ps from data

Select the Advanced Encryption standard (AES) block shifting

$$Chs = \int_{i=1}^{size(Nat)} n\varnothing \left( Chs- \rightarrow internalkeyK(i) \right)$$

Step 2:   Select a circular shifting to shift block.

Select encryption key Epk.

Endpoint key Epk $\rightarrow$ initiate public key each level R

$$Epk = \int_{n}^{1} rCr(R)$$

Add to keyset ks.

$$Epk = \sum_{n}^{1} k \left( Key \in K- \rightarrow starvationpoints \right) \cup Ekey$$

Step 3:   Return encryption stage of starvation point Epk

Epk=Eps++;

End

Stop.

---

Cancel IBEE Program in the IBEE scenario, in which the players involved in the show are key officials, the withdrawn user is not withdrawn. The system has also been updated with the key authority to decode and release major update material. These keywords will not only be revoked but will be updated for users. This way, if the user is withdrawn as a member or he/she is found to be malicious, users will not make his or her important compromise.

So IBBE has many receiver configurations of Blockchain, which can be a good candidate for providing security and privacy of your transaction data. This application allows you to demonstrate active records as a member of blockchains with authentication. The Revoked

Identity-Based Broadcasting Encryption (RIBBE) method can work efficiently in the case of Blockchain as it is highly efficient in terms of complex forecast communication.

## 12.12  Peer Verification on Exponential Session

End-to-end verification based on the authenticity performed by Blockchain personal messages authentication. It can use Blockchain for secure communication between the parties.

---

Input: preprocessed data Ps, Exponential session time ET
Output: output encrypted text
Start

Step 1: Two exponential prime numbers P and Q is used to generate max confidence

Step 2: Process the session-based data encrypt using a two-factor key
    If (the prime factor p ≠ q such that. p & q) key factor
    {
      Generate on time session key → Sk
      Compute n= p × q;
}end if

Step 3: Calculate the intensity of data
If ( $d(n) = (p-1)(q-1)$.) factors of exp value e
    {
      The exponential integer value be chosen 1<e → Ps as e
      User A possess the message m to encrypt B → A
   Whether A be message decrypts, the authentication followed to user B User
   A attained to Get the secure level public keys (nA, eA).
      Update on session T → Ps
    }

Step 4: compute the terms message at the regular interval [0, nA − 1].
Comute the random point of selection k, 1 < k <nA, such that gcd(k, nA)) = 1.
    if (c1 = k eA mod nA.) and (c2 = meA k mod nA)
      {
      Transfer the ciphertext request to user A as (c1,c2).
      Return on state session T
      }
   End if
End if
Stop

---

The Elliptic Curve Diffie-Hellman Merkle (ECDHM) addresses peer requests end on authentication which is required by attended participants, and this shared secret is used by each other to obtain the anonymous trans address. This address can only be exposed if they have a role to play in creating these addresses. By the transaction, the repeated request was verified by the privacy logs under the defined access level of data privacy concerns.

## 12.13  Verified Authenticity Key Logs

Confirm data through the audit verification commission with policies to manage cloud service integrity. Data requests are verified by accessing the keylog provided by the owner company to submit to the cloud service provider. Cloud service providers enrich archived logs validation. The key is providing the right authentication to see if it is the right access. The key is validated after the file is allowed to decrypt the data.

In the above algorithm, to provide data, it converts the hunger verification key verification form to access owners. Permission is granted to the data owner, the data provider. Integrates tons of verifiable access security. Finally, the authentication verification key provides permission to decrypt the verification data.

---

**ALGORITHM**

Input: Number of encrypted Clair text, verify the security key.
Start

Step 1: Compute the key verification to obtain user request R $\rightarrow$ Req.
Step 2: Compute the key validation

    If Req. Type R==Enter key Then.

        Verify the log to access data $\rightarrow$ TPa.

      Transmit Req R $\rightarrow$ key observe right access

    Else if Req.Type==no match Then
Step 3: Authenticate through third-party auditing

      If True then

        Return service to decrypt data R.

      End

    End
Step 4: verified access rights to return encrypted text to decrypt

      R $\rightarrow$ reverse decryption;
Stop.

---

## 12.14  Result and Discussion

To propose a RSCRE using circular shift round random padding key for decentralized cloud environment (CSR2PK). With a high degree of security to keep facts accurate, time calculation of user roles has had a major impact on access to cloud environment security. The result is that the client-server participant response verification verifies the security level of the test in access control privacy. A test case will be created by configuring a Microsoft cloud custom configuration tool intended to work with a SQL Server database called a Pentium processor to become i3. It has permissions for private and public users and access users. Users can access thousands of trust files and demonstrate the high impact of the rating on privacy issues. Programs to process the parameters are shown in the tables below (Tables 12.1–12.3).

**TABLE 12.1**

Parameters and Its Value Processed

| Parameter | Name |
|---|---|
| Cloud service | Amazon web service |
| Storage model | S3 glazier |
| Tool used | Accord visual framework |
| Number of users/ file type | 1000/ content file |

**TABLE 12.2**

Methods and Its Security Performance

| Methods | Security Performance in Ms |
|---|---|
| sub-Enc | 68.3 |
| Ceaser cipher | 78.4 |
| SHA | 84.9 |
| Tri-DES | 92.6 |
| S2OPE | 95.2 |
| RSCRE | 96.4 |

**TABLE 12.3**

Methods and Its Frequent Occurrence

| Methods | A frequent occurrence in % | | |
|---|---|---|---|
| | 25 MB | 50 MB | 75 MB |
| sub-Enc | 9.3 | 72.4 | 82.7 |
| Ceasercipher | 7.4 | 54.4 | 62.4 |
| SHA | 6.7 | 41.8 | 53.3 |
| Tri-DES | 4.1 | 22.8 | 32.5 |
| S2OPE | 3.6 | 15.1 | 17.8 |
| RSCRE | 3.2 | 14.2 | 15.9 |

Tables 12.1–12.2 qualify the security concerns parameters used in such structures. The diagram below shows the different performances of the traditional comparative test analyses.

Figure 12.3, shows the ability to process execution between encryption and delete using 27.8 ms as well DES vortex principle provides good alternatives if the replacement time modules are separated. This implementation had significantly improved performance over the traditional method.

Security presentation can be examined to finish the total number of susceptibilities of attacks approved out by an un-authenticated procedure that indications to file decryption by receiving the plain text.

Figure 12.4, demonstrates the relative examination of security presentation well to dissimilar approaches, and this enables inordinate presentation with more efficiency than preceding approaches.

**Analysis of Execution state encryption and decryption**

| | sub-Enc | ceaser chipher | SHA | Tri-DES | S2OPE | RSCRE |
|---|---|---|---|---|---|---|
| Encryption | 68.1 | 55.4 | 45.1 | 27.4 | 15.6 | 14.2 |
| Decryption | 82.4 | 64.4 | 51.8 | 32.8 | 20.1 | 18.6 |

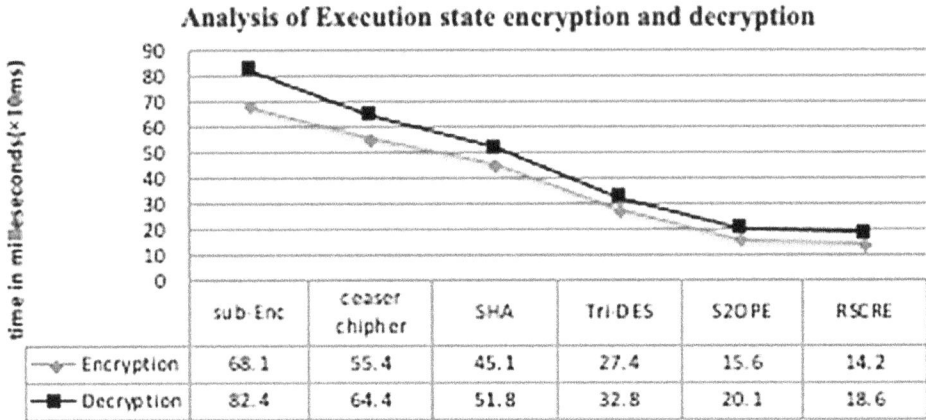

**FIGURE 12.3**
Comparison of execution efficiency.

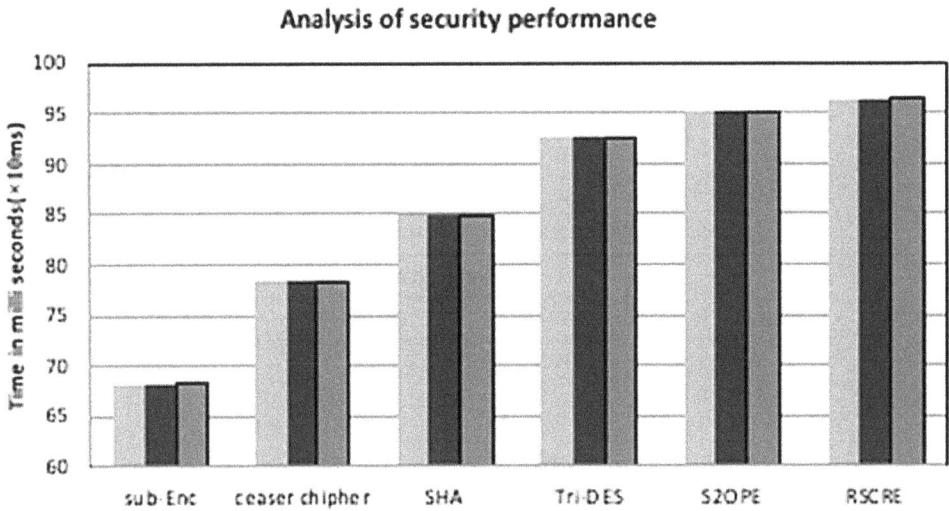

**Analysis of security performance**

**FIGURE 12.4**
Comparison of security analysis efficiency.

$$\text{Time complexity}\,(Ts) = \frac{\text{Total number of blocks per bits} \times \text{two phase encryption}}{\text{time taken}\,(s)}$$

Figure 12.5 shows the processing time for different file sizes to run differently, the time as well as the DES crypto policy. The zipper holds unnecessary time tracks because the volume separation by the DES package creates complex time. This implementation had significantly improved performance over the traditional method.

$$\text{Frequent occurrence state}\,(FS) = \frac{\text{Repeated block of the cipher}}{\text{Total number of cipher block occurrence}}$$

## Analysis of Time complexity

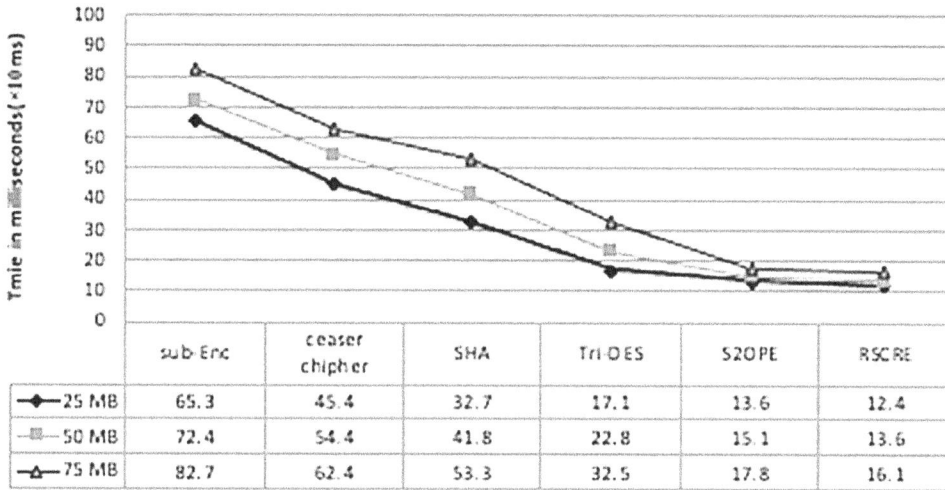

| | sub-Enc | ceaser chipher | SHA | Tri-DES | S2OPE | RSCRE |
|---|---|---|---|---|---|---|
| 25 MB | 65.3 | 45.4 | 32.7 | 17.1 | 13.6 | 12.4 |
| 50 MB | 72.4 | 54.4 | 41.8 | 22.8 | 15.1 | 13.6 |
| 75 MB | 82.7 | 62.4 | 53.3 | 32.5 | 17.8 | 16.1 |

**FIGURE 12.5**
Comparison of time complexity.

## Analysis frequent occurence

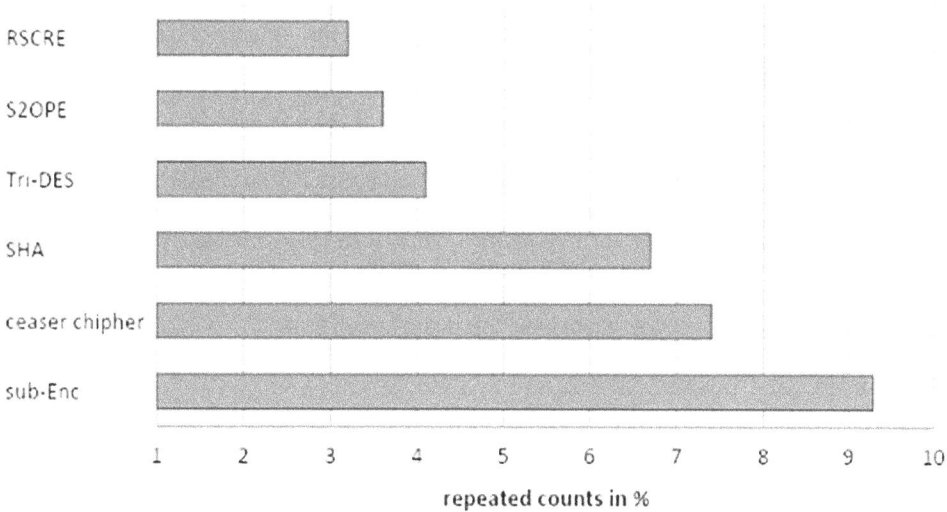

repeated counts in %

**FIGURE 12.6**
Comparison of frequent occurrence.

Figure 12.6 shows that, due to the encryption of encryption policies that are comparable in different ways to the cryptography policy, ciphertexts are a frequent occurrence, more often than not, and a repeating device.

Our activation method produces 3.5% more active unwanted frequency states than previous methods.

## 12.15  Conclusion

Considering that Blockchain is a very powerful tool for applying data security and integrity on cloud platforms. We take security measures in the updated Blockchain type user ownership policy along with our data protection application. To propose a RSCRE using circular shift round random padding key for decentralized cloud environment (CSR2PK) to improve the cloud security. The proposed system produces 96.4% higher security performance than previous methods, lasting integrated with PeerVerification Security in Security Parade. Blockchain on the cloud is still exploring the availability of concrete data and improve security in integrity applications.

## References

1. X. Wang, X. Lai, D. Feng (2005) Cryptanalysis of the Hash Functions MD4 and RIPEMD. *Advances in Eurocrypt*, 3494: 1–18.
2. G. Zyskind, O. Nathan, and A. Pentland, "Decentralizing Privacy: Using Blockchain to Protect Personal Data," in *2015 IEEE Security and Privacy Workshop*, San Jose, CA, 2015, pp. 180–184.
3. C. Fromknecht, D. Velicanu, and S. Yakoubov, "A Decentralized Public Key Infrastructure with Identity Retention, in *IACR Cryptology ePrint Archive*, 2014, pp. 803.
4. T. ElGamal, "A public-key cryptosystem and a signature scheme based on discrete logarithms," *IEEE Transactions on Information Theory*, vol. 31, no. 4, 1985, pp. 469–472.
5. X. Li, P. Jiang, T. Chen, X. Luo, and Q. Wen, "A survey on the security of blockchain systems," *Future Generation Computer Systems*, arXiv:1802.06993, 2018.
6. T. T. Dinh, R. Liu, M. Zhang, G. Chen, B. C. Ooi, and J. Wang (2018). Untangling blockchain: A data processing view of blockchain systems. *IEEE Transactions on Knowledge and Data Engineering*,30(7), 1366–1385. doi:10.1109/tkde.2017.2781227.
7. C. Allen, A. Brock, V. Buterin, J. Callas, D. Dorje, C. Lundkvist, P. Kravchenko, J. Nelson, D. Reed, M. Sabadello, G. Slepak, N. Thorp, and H. T. Wood, "Decentralized Public Key Infrastructure," *A White Paper from Rebooting the Web of Trust*, 2016, [online] Available: https://danubetech.com/download/dpki.pdf, (accessed February 13, 2018).
8. M. Niranjanamurthy, B. N. Nithya, and S. Jagannatha, "Analysis of Blockchain technology: pros, cons and SWOT," *Cluster Computing*, pp. 1–15, March 2018.
9. K. Christidis and M. Devetsikiotis, "Blockchains and smart contracts for the Internet of Things," *IEEE Access*, vol. 4, pp. 2292–2303, 2016.
10. S. Gan, "An IoT simulator in NS3 and a key-based authentication architecture for IoT devices using blockchain," M.S. thesis, Indian Inst. Technol. at Kanpur, Kanpur, India, 2017. Accessed: Feb. 13, 2018. [Online]. Available: https://security.cse.iitk.ac.in/node/240.
11. T. T. Dinh et al., "BLOCKBENCH: A Framework for Analyzing Private Blockchains," in *Proceedings ACM International Conferences Management Data (SIGMOD)*, New York, NY, USA, 2017, pp. 1085–1100.
12. K. Kwiat, and L. Njilla, "ProvChain: A Blockchain-Based Data Provenance Architecture in Cloud Environment with Enhanced Privacy and Availability," *17th IEEE/ACM International Symposium on Cluster, Cloud and Grid Computing (CCGRID)*, pp. 468–477, May 2017.
13. W. Wang, D. T. Hoang, P. Hu, Z. Xiong, D. Niyato, P. Wang, Y. Wen, and D. I. Kim, "A survey on consensus mechanisms and mining strategy management in blockchain networks," 2018, arXiv:1805.02707. [Online]. Available: https://arxiv.org/abs/1805.02707.

14. A. Regenscheid, R. Perlner, S.-J. Chang, J. Kelsey, M. Nandi, and S. Paul, "Status report on the first round of the SHA-3 cryptographic hash algorithm competition," Inf. Technol. Lab., Nat. Inst. Standards Technol., Gaithersburg, MD, USA, Tech. Rep. NISTIR 7620, 2009.

15. K. Gongsheng. Advances on secure authentication and trusted admission protocols for cloud computing, *Journal of Henan University*, 2017.

16. Y. Zhu, G. H. Gan, D. Deng (2016) Security research in key technologies of Blockchain. *Information Security Research*, 12, 1090–1097.

17. M. Andrychowicz et al., "Fair Twoparty Computations Via Bitcoin Deposits," in *Financial Cryptography and Data Security. FC 2014. Lecture Notes in Computer Science*, vol 8438. Springer, Berlin, Heidelberg. doi: 10.1007/978-3-662-44774-1_8, pp. 105–121.

18. Y. Li et al., "Fuzzy identity-based data integrity auditing for reliable cloud storage systems," *IEEE Transaction Dependable and Secure Computing*, 2016. doi: 10.1109/TDSC.2017.2662216.

19. Z. Guan et al., "Achieving efficient and secure data acquisition for cloud-supported Internet of Things in smart grid," *IEEE Internet of Things Journal*, vol. 4, no. 6, December 2017, pp. 1934–1944.

20. B. Shahzad, and J. Crowcroft (2019). Trustworthy electronic voting using adjusted blockchain technology. *IEEE Access,*7, 24477–24488. doi:10.1109/access.2019.2895670.

21. K. Fan, Y. Ren, Y. Wang, H. Li, and Y. Yang (2018). Blockchain-based efficient privacy preserving and data sharing scheme of content-centric network in 5G. *IET Communications*, 12(5), 527–532. doi:10.1049/iet-com.2017.0619.

22. A. E. Kosba, A. Miller, E. Shi, Z. Wen, and C. Papamanthou, "Hawk: The blockchain model of cryptography and privacy-preserving smart contracts," in *IEEE Symposium on Security and Privacy, SP* 2016, San Jose, CA, USA, May 22–26, 2016, pp. 839–858

23. S. Hu, C. Cai, Q. Wang, C. Wang, X. Luo, and K. Ren (2019). "Searching an encrypted cloud meets blockchain: A decentralized, reliable, and fair realization," *IEEE INFOCOM*, 2018, pp. 792–800.

24. L. Chen, W.-K. Lee, C.-C. Chang, K.-K. R. Choo, and N. Zhang "Blockchain based searchable encryption for electronic health record sharing," *Future Generation Computer Systems (FGCS)*, 95, pp. 420–429.

# 13

## Application of Exact Barrier-Penalty Function for Developing Privacy Mechanisms in Cyber-Physical Systems

**Manas Kumar Yogi**

*Pragati Engineering College (A), Peddapuram Town, India*

**Ardhani Satya Narayana Chakravarthy**

*JNTUK-University College of Engineering, Kakinada, India*

## CONTENTS

## 13.1 Introduction

Cyber-physical systems are becoming an indispensable part of modern-day lives in the current era. Among the multi-operations of cyber-physical systems (CPS), starting from, smart transportation systems, smart homes smart urban communities, smart energy systems, is that data gathered from individuals or elements serve an indispensable part of dynamic and control hidden the systems' operation (R. Jia et al., 2017). The advancement in CPS releases an emerging area of multi-disciplinary participation, connecting software engineering and control hypothesis with a few designing regions, innate sciences, and medication (S. Ghayyur et al., 2018). Progressively, CPSs are further developing the execution, usefulness, and productivity of energy in the control of physical entities. Specialists and professionals are planning and developing robust and dynamic prototypes such as independent vehicles with more significant levels of robotization and network. Also, the medical care area is creating novel clinical applications to all the more likely help and treat patients, including independent implantable gadgets and framework structures for checking clinical patients or patients residing in remote locations. Among other important CPS

**TABLE 13.1**

Popular Privacy Preservation Schemes Pertaining to CPS

| Sl. No. | Approach | Strength | Limitations |
|---|---|---|---|
| 1 | Cryptographic methods | High level of integrity, safe, trustworthy, computations are possible on encrypted datasets | Increase in communication complexity, Scalability is an issue |
| 2 | Data mining methods | High level of privacy, efficiency | Linking attacks cannot be faced |
| 3 | Statistical methods | High level of privacy, Improvement in data analytics due to data projection | Degradation in data utility due to disturbance in data, difficulty in balancing between privacy and data correction |
| 4 | Authentication based models | Third party authentication is possible, degree of data integrity can be controlled without much effort | Poor system reliability, processing overhead, data storage cost is also high |
| 5 | Blockchain based methods | No single point of failure High degree of data transparency and traceability High level of user confidentiality | High operation and customization costs Blockchain literacy Adaptability challenges Scalability challenges |
| 6 | Machine learning and deep learning-based methods | Easy application of randomization techniques, learns from large datasets, Data analytics becomes easy | Cost of parameter setting becomes high, Data utility becomes costly, overhead in computational resources |

applications include mechanical control systems for assembling and interaction plants, mechanical technology, control systems in basic frameworks offering fundamental types of assistance to networks, and independent military safeguard rockets (S. Desai et al., 2018). Thinking about the promising turns of events and the basic utilizations of CPSs, government organizations and mechanical associations respect the exploration endeavors in CPSs as a need (Github, 2020). Table 13.1 below enumerates the different approaches for privacy preservation in CPS and brings out their relative strengths and limitations.

## 13.2 Related Work

In recent times, database aspects are considered while devising privacy preservation techniques. Majorly, two groups have been identified to work on this research domain. The first one is using anonymization techniques and the second one is applying various models of differential privacy (Heng Ye et al., 2020). In group one, the state of work reached up to robust techniques ranging from k-anonymity to l-diversity and most recently up to t-closeness methods. In group two state of work in recent times reached the mechanisms of membership privacy and differential identifiability (Jian Xu et al., 2020). Many researchers have also worked on the usage of synthetic data in differential privacy techniques which have become quite popular (Farokhi & Farhad, 2020). But involving synthetic data principles in the design of privacy preservation mechanisms is complex to implement. For applications involving big data becomes rigorous. That being said, synthetic data usage along with variants of differential privacy methods with gain momentum for privacy preservation in CPS in next few years (Keshk et al., 2019). In most of the differential privacy techniques, random noise value is added to the data so that attackers cannot identify the records in specific and the degree to which they can attack the user's privacy takes a backseat. Various mechanisms have been developed to add noise. In these Epsilon differential privacy methods, noise is added to improve the data utility rate and the error rate is minimized. In this scheme, data protection is achieved by mechanisms of perturbation. Gaussian mechanism, Laplace mechanism, Exponential mechanism are used to add noise to the original data (Degue, 2021). The degree of noise to be added is directly proportional to the measure of global sensitivity and the amount of privacy budget desired. In Laplace mechanism, the degree of noise to be added is determined with the Laplacian function, and each data coordinate is perturbed with the help of the Laplacian noise which is deduced from the LM distribution (Bhaskar et al., 2020). The scale of noise being added is gauged from the sensitivity factor of the differential privacy function. For the researchers, selecting the optimal value of Epsilon is still a challenge since it affects the trade-off between privacy and data utility (Hou et al., 2020). A smaller value of Epsilon may increase privacy level but degrades the level of data utility. Few researchers have worked on techniques to identify the optimal Epsilon value. Among them, notable work is related to criteria like largest affordable value and factor of success probability. Nevertheless, choosing the Epsilon value in optimal fashion still remains a goal to achieve (Qu et al., 2020). Many researchers have also demonstrated strategies related to information-theoretic privacy employing techniques like Shannon entropy which limits guessing attacks on user privacy. But compared to differential privacy mechanisms, they don't possess a tight bound on the malicious attackers targeting the privacy of users (Jiang et al., 2021).

## 13.3 Problem Formulation

In this part of the chapter, the privacy problem is formulated as a constrained nonlinear optimization problem where privacy loss has to be minimized.

In our objective of minimizing the privacy loss, sources emitting private data might be required to remove personally identifiable information (PII) from the dataset. Let $x_j$ be the amount of PII to be removed at source j. Our proposed mechanism tries to minimize the privacy loss:

$$\text{Minimize} \sum_{i=1}^{n} fj(xj) \tag{13.1}$$

Subject to:

$$\sum_{j=1}^{n} aijxj \geq b_i \text{ where i values ranges from 1, 2... and so on, up to m.}$$

$$0 \leq x_j \leq u_j \ (j = 1, 2, ..., n),$$

Below notations are used to represent the various participating features.

$f_j(x_j)$ denotes the Cost of removing $xj$ amount of PII at source $j$,

$b_i$ represents the desired improvement in privacy loss at functional unit $i$ in the CPS,

$a_{ij}$ denotes Quality of privacy loss, at functional unit $i$ in the system, caused by removing PII at source $j$,

$u_j$ represents the maximum degree of PII that can be removed at source $j$.

## 13.4 Proposed Mechanism

It is observed that the penalty function is playing a crucial role in determining multiple constrained optimization issues in the factory design domains. It is conventionally developed to work out nonlinear functions by integrating some degree of penalty factor or barrier factors when applied to the limitations related to the function which needs to be minimized. Then it can be optimized by some unbounded or bounded optimization mechanisms or sequential quadratic programming (SQP) approaches. Irrespective of the applied technique, the penalty function largely relies on a modest parameter $\varepsilon$. As the value of $\varepsilon$ approaches 0; the minimum value pertaining toward the penalty function merges to the original problem's minimizer. By introducing the exact penalty function properties, a minuscule value of the privacy budget, $\varepsilon$ acts a minimizer with respect to the original problem. For our case, the privacy loss will be denoted by $\varepsilon$, which has to be minimized by using the exact penalty function.

In principle, our problem of minimizing the privacy loss can be expressed as

$$\text{Minimize } f\left(x^{\varepsilon}\right) \text{ subject to } x \le u_j \text{ and } x > b_j \tag{13.2}$$

Consider Pen($x$), which denotes a penalty for infeasible solutions, represented as:

$$\text{Pen}\left(x\right) = +\infty \text{ if } x \text{ becomes infeasible solution}\left(x > u_j \text{ or } x > b_j\right) \tag{13.3}$$

$= 0$ if $x$ results in a feasible solution (that is, $b_j \le x \le u_j$).

Consequently, the constrained optimization problem can be rehashed in an unbounded state as

$$\text{Minimize}\left\{f\left(x^{\varepsilon}\right) + \text{Pen}\left(x\right)\right\} \tag{13.4}$$

It can be deduced that for getting a feasible answer to the bounded problem, the proposed objective function is aligned to the original objective function. Needless to say, the exact penalty function acts as a minimizer function to control the privacy loss. Though the privacy loss minimization function is represented in terms of penalties, the implementation of the method is rendered difficult due to the presence of $+\infty$. Efforts can be applied to reduce this obstacle by applying approximation methods to the penalty term. Now, substitute the penalty factor P($x$) by factors of 2P($x$), 3P($x$), 4P($x$), or, in general, $r$P($x$) for $r > 1$, the penalty escalates for any unbounded boundary. When the value of $r$ increases gradually, the overall penalty function value for the original objective function also increases as shown below:

Minimize $\{f\left(x^{\varepsilon}\right) + r\,\text{P}(x)\}$ is brought near to the feasible region.

Consider a situation where $r > 1$, then the penalty problem can be solved by using $f\left(x^{\varepsilon}\right) = 1$, in which $r\,(1 - f(x^{\varepsilon}))^2$ constitutes the penalty term. Now, the penalty problem of minimizing

$\{f(x^{\varepsilon}) + r\,\text{P}(x)\}$ reduces to:

$$\text{Minimize}\left\{f\left(x^{\varepsilon}\right) + r\left(1 - f\left(x^{\varepsilon}\right)\right)^2\right\}. \tag{13.5}$$

Determining the first derivative for this objective function and equating it to zero gives

$$d\left(f\left(x^{\varepsilon}\right)\right) - 2r\left(1 - d\left(f\left(x^{\varepsilon}\right)\right)\right) = 0 \tag{13.6}$$

$$d\left(f\left(x^{\varepsilon}\right)\right) = 2r\left(1 - d\left(f\left(x^{\varepsilon}\right)\right)\right) \tag{13.7}$$

Now applying the barrier methods on Equation 13.7, the absolute values for $f(x^{\varepsilon})$ can be obtained.

It can be easily observed that the barrier term $1/(f(x^{\varepsilon}) - b_j)^2$ tends to be infinite as $x$ moves to the range of the constraint, where $f(x^{\varepsilon}) = b_j$, if the function originates with a feasible solution initially, the minimization function will not allow it to cross the boundary, if it passes the boundary then it becomes infeasible. As value of r reaches a high limit, the barrier factor reduces near the boundary and the terms in Equation 13.7 start appearing like the penalty function with $P(x) = 0$ where the value of $x$ is feasible and $P(x)$ becomes infinity when $x$ tends to be infeasible.

If $\alpha$ denotes the optimal solution for Equation 13.7, then $r \geq \varepsilon$ solves the original objective function.

Hence it can be observed that the privacy loss at a node $j$ can be reduced to a great extent by modeling the functionality of a node in CPS as a constrained nonlinear function. It is known that a cyber-physical system environment is dynamic in state and its functionality can be modeled with respect to the time series equation too (Bi et al., 2020). This feature adds value to our proposed mechanism because by applying the exact penalty function and barrier methods our approach fits directly into the CPS ecosystem where the focus is on the privacy loss factor.

---

## 13.5 Experimental Results

In this section, the validation of applicability of the proposed model with an example study of a smart health management system where the privacy of the patients is presented concisely.

Quite possibly the most alluring operation of interlacing the physical and digital world is medical and healthcare frameworks. This association has a colossal impact on CPSs, and it brings about multiple applications related to generic healthcare, for example, real-time health monitoring, workout schedules, distant health monitoring. Another potential application of this association is treatment medication and from places remote in nature. In the same sense, health-related records storage using enormous data, and implementing the concerned realizations of information analytics reviews for improved analysis of illness at the initial stage is also being worked on phase. It was observed during our study that in CPSs related to healthcare, technologies like the ultra-narrow band (UNB), and 4G long-term evolution (LTE), low-power wide area (LPWA) techniques are deployed for communication purposes. Most of these popular and robust standards send health data in real-time by introducing minimum time lag. The inherent health records have data that includes definite arrangements for real-time e-health verified data, and these arrangements ought to be secured with an unperturbed degree of privacy control since they are straightforwardly connected with a user's personal activities (Jairo et al., 2020). For example, consider the diagnosis of any certain illness, ending date of health insurance, a certain degree of level of glucose in the body, date of appointment from a doctor and so forth on the off chance that any intruder gains access to sensitive data, it can straightforwardly or indirectly have an impact on the existence of the patient. With the swift expansion of remote gadgets in our day-to-day lives, the way people handle their health-related aspects keeps changing dynamically. Data related to health is being accounted for to databases or medical experts in real-time to monitor client conduct and movement, like, heart rate information in various conditions, circulatory strain; count of walk steps can be shared with insurance companies, hospital authorities. Nonetheless, the divulgence of data which seems unnecessary can lead to extreme concerns of privacy. During the context of data sharing related to health records, two aspects are considered as the primary goal:

1. Utility (handiness of data) and
2. Privacy (divulgence of minimum private data).

Wearable medical gadgets act as prominent sources of real-time health data; these gadgets are non-invasive, and autonomous gadgets intended to play out any specific medical capacity, for example, monitoring data related to a person's health. The integral indications of patients, for example, blood oxygen level, blood pressure, body fat, heart rate and state of respiration are monitored continuously and observed to know about future undesirable ailments. Also, athletes utilize various types of wearable medical gadgets in request to monitor their, pace, speed and heart rate, calorie burn in the course of exercise and the concerned coaches gets a specialized report on them. Now this data includes specific arrangements, which may give important inference data regarding health of any individual. Notwithstanding, imagine that any patient's private data is made under the control of an attacker, and then the concerned user may have to experience serious health problems.

In the above Figures 13.1 and 13.2, (1) represents the useful information collected from smart wearable devices from the users which are stored in a central database; (2) denotes the data collected from hospital records are also directed toward the central database;

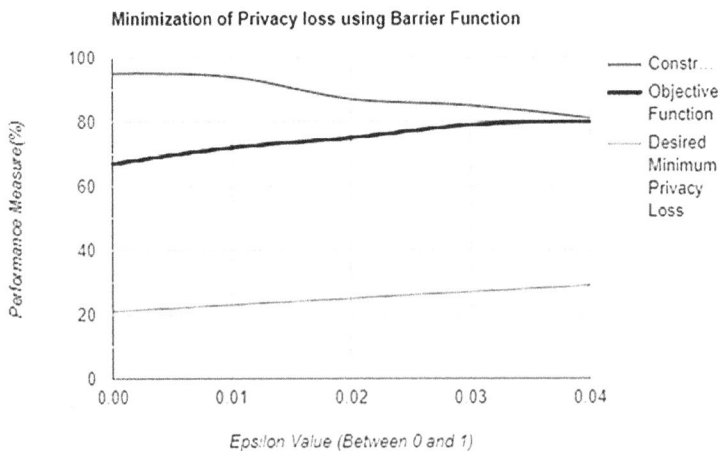

FIGURE 13.1
Minimization of privacy loss using the barrier function.

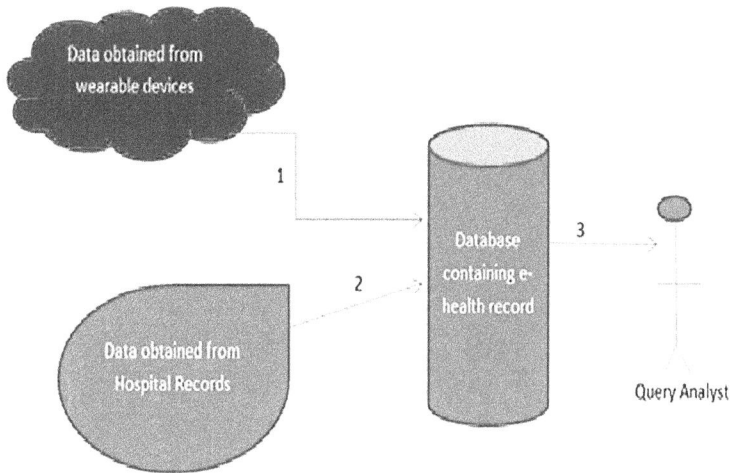

**FIGURE 13.2**
Illustrative mechanism for Smart Health Management System.

(3) represents the proposed mechanism which applies the exact penalty-barrier function to the health records and removes the sensitive PII from the dataset so that the query analyst gets only the minimum information which can be used for further processing.

In this study, a cancer patient's dataset has been downloaded from Kaggle which had 1000 rows and over 20 features based on which their treatment was undertaken. The following Figures 13.3–13.5 show a sample of the dataset on which the proposed mechanism is applied (Table 13.2).

Over a few years, the enticing practice of medical clinics embracing the electronic method of saving patient-sensitive data has increased drastically. This approach is popularly called as e-health technique which integrates advanced communication mechanisms. This health information includes PII, like blood pressure level, disease pre-existence, heart rate monitoring on either weekly or monthly, medical symptoms, date of birth etc. This sensitive information is extremely private which gets stored in datasets and shall not be divulged to outsiders except for the doctor and patient (Table 13.3).

As it can be observed from above, the optimal value at $\varepsilon = 0.11$ brings down the privacy loss to nearly 15% compared to $\varepsilon = 0.50$. Hence it is advocated that the privacy engineers work with $\varepsilon = 0.11$ so as to get more than satisfactory results with the exact penalty-barrier function method. Values less than 0.11 produces infeasible results and also computational overhead increases.

Ordinarily, the sensitive medical records are protected by utilizing cloud computing mechanisms as well as anonymizing operations amid information cleaning and arrangement. In the cloud computing approach, attributes like the keys along with a few other types of sensitive information features are disguised; subsequently, for the purpose of mining, a dataset is constructed which is in turn protected in nature.

Nevertheless, the masked records are prone to be exposed with respect to few PII when they are investigated and blended in with various other capabilities.

As evident from the graph above, the proposed mechanism outperforms the other two popular methods of K-means clustering-based DP mechanism and Laplace mechanism by nearly 2.5% in terms of reduction in the privacy loss. In a generic sense, the data utility loss is also minimal which can be accepted by most of the users.

| Patient Id | Age | Gender | Air Pollution | Alcohol use | Dust Allergy | OccuPational Hazards | Genetic Risk | chronic Lung Disease | Balanced Diet | Obesity | Smoking | Passive Smoker | Chest Pain | Coughing of Blood | Fatigue | Weight Loss | Shortnes s of Breath | Wheezing | Swallowi ng Difficulty | Clubbing of Finger Nails | Frequent Cold | Dry Cough | Scoring | Level |
|---|---|---|---|---|---|---|---|---|---|---|---|---|---|---|---|---|---|---|---|---|---|---|---|---|
| P1 | 33 | 1 | 2 | 4 | 5 | 4 | 3 | 2 | 2 | 4 | 3 | 2 | 2 | 4 | 3 | 4 | 2 | 2 | 3 | 1 | 2 | 3 | 4 | Low |
| P23 | 17 | 1 | 3 | 1 | 5 | 3 | 4 | 2 | 2 | 2 | 2 | 4 | 2 | 3 | 1 | 3 | 7 | 8 | 6 | 2 | 1 | 1 | 2 | Medium |
| P200 | 35 | 1 | 4 | 5 | 6 | 5 | 5 | 4 | 6 | 7 | 2 | 3 | 4 | 8 | 8 | 7 | 9 | 2 | 1 | 4 | 6 | 7 | 2 | High |
| P2000 | 37 | 1 | 7 | 7 | 7 | 7 | 6 | 7 | 7 | 7 | 7 | 7 | 7 | 8 | 4 | 2 | 3 | 1 | 4 | 5 | 6 | 7 | 5 | High |
| P221 | 45 | 1 | 6 | 8 | 7 | 7 | 7 | 6 | 7 | 7 | 8 | 7 | 7 | 9 | 3 | 2 | 4 | 1 | 4 | 2 | 4 | 2 | 3 | High |
| P222 | 35 | 1 | 4 | 5 | 6 | 5 | 5 | 4 | 6 | 7 | 2 | 3 | 4 | 8 | 8 | 7 | 3 | 2 | 1 | 4 | 6 | 7 | 2 | High |
| P223 | 52 | 2 | 2 | 4 | 5 | 4 | 3 | 2 | 2 | 4 | 3 | 2 | 2 | 4 | 3 | 4 | 2 | 2 | 3 | 1 | 2 | 3 | 4 | Low |
| P224 | 28 | 2 | 3 | 1 | 4 | 3 | 2 | 3 | 4 | 3 | 1 | 4 | 3 | 1 | 3 | 2 | 2 | 4 | 2 | 2 | 3 | 4 | 3 | Low |
| P225 | 35 | 2 | 4 | 5 | 6 | 5 | 6 | 5 | 5 | 5 | 6 | 6 | 6 | 5 | 1 | 4 | 3 | 2 | 4 | 6 | 2 | 4 | 1 | Medium |
| P226 | 45 | 1 | 2 | 3 | 4 | 2 | 4 | 3 | 3 | 3 | 2 | 3 | 4 | 4 | 1 | 2 | 4 | 5 | 5 | 4 | 2 | 1 | 5 | Medium |
| P227 | 44 | 1 | 5 | 7 | 7 | 7 | 7 | 6 | 7 | 7 | 7 | 8 | 7 | 7 | 5 | 3 | 2 | 7 | 8 | 2 | 4 | 5 | 3 | High |
| P228 | 54 | 2 | 5 | 8 | 7 | 7 | 7 | 6 | 7 | 7 | 7 | 8 | 7 | 7 | 9 | 6 | 5 | 7 | 2 | 4 | 3 | 1 | 4 | High |
| P229 | 35 | 2 | 4 | 5 | 6 | 6 | 5 | 4 | 6 | 6 | 6 | 6 | 6 | 6 | 5 | 3 | 2 | 4 | 3 | 1 | 7 | 5 | 6 | Medium |
| P21 | 34 | 1 | 5 | 7 | 7 | 7 | 6 | 7 | 7 | 7 | 7 | 7 | 7 | 8 | 4 | 2 | 3 | 1 | 4 | 5 | 6 | 7 | 5 | High |
| P220 | 27 | 2 | 3 | 1 | 4 | 2 | 3 | 2 | 3 | 3 | 2 | 2 | 4 | 2 | 2 | 3 | 4 | 1 | 5 | 2 | 6 | | 2 | Low |
| P211 | 73 | 1 | 5 | 6 | 6 | 5 | 6 | 5 | 6 | 5 | 8 | 5 | 5 | 5 | 4 | 3 | 6 | 2 | 1 | 2 | 1 | 6 | 2 | Medium |
| P212 | 17 | 1 | 3 | 1 | 5 | 3 | 4 | 2 | 2 | 2 | 2 | 4 | 2 | 3 | 1 | 3 | 7 | 8 | 6 | 2 | 1 | 1 | 2 | Medium |
| P213 | 34 | 1 | 5 | 7 | 7 | 7 | 6 | 7 | 7 | 7 | 7 | 7 | 7 | 8 | 4 | 2 | 3 | 1 | 4 | 5 | 6 | 7 | 5 | High |
| P214 | 35 | 1 | 6 | 7 | 7 | 7 | 7 | 7 | 6 | 7 | 7 | 7 | 7 | 7 | 8 | 5 | 7 | 5 | 7 | 8 | 7 | 6 | 2 | High |
| P215 | 34 | 1 | 2 | 4 | 5 | 6 | 5 | 5 | 4 | 6 | 5 | 4 | 6 | 5 | 5 | 3 | 2 | 1 | 4 | 7 | 2 | 1 | 6 | Medium |
| P216 | 24 | 1 | 6 | 8 | 7 | 7 | 6 | 7 | 7 | 3 | 8 | 7 | 9 | 6 | 5 | 2 | 5 | 2 | 3 | 2 | 1 | 7 | 6 | High |
| P217 | 53 | 2 | 4 | 5 | 6 | 5 | 5 | 4 | 6 | 7 | 2 | 3 | 4 | 8 | 7 | 9 | 2 | 1 | 4 | 6 | 7 | | 2 | High |
| P218 | 62 | 1 | 6 | 8 | 7 | 7 | 7 | 6 | 7 | 7 | 8 | 7 | 7 | 9 | 3 | 2 | 4 | 1 | 4 | 2 | 4 | 2 | 3 | High |
| P219 | 25 | 2 | 5 | 7 | 7 | 7 | 7 | 6 | 7 | 7 | 7 | 7 | 7 | 7 | 2 | 7 | 6 | 7 | 6 | 7 | 2 | 3 | 1 | High |
| P22 | 35 | 1 | 5 | 7 | 7 | 7 | 7 | 7 | 6 | 7 | 7 | 7 | 7 | 8 | 5 | 7 | 5 | 7 | 8 | 7 | 6 | | 2 | High |
| P220 | 65 | 1 | 6 | 8 | 7 | 7 | 7 | 6 | 2 | 4 | 1 | 2 | 4 | 3 | 2 | 7 | 6 | 5 | 1 | 5 | 3 | 4 | 2 | Medium |
| P221 | 38 | 2 | 2 | 1 | 5 | 3 | 2 | 3 | 2 | 4 | 1 | 4 | 2 | 4 | 6 | 7 | 2 | 5 | 8 | 1 | 3 | 2 | 3 | Medium |
| P222 | 13 | 1 | 3 | 2 | 4 | 2 | 3 | 2 | 3 | 3 | 2 | 2 | 3 | 3 | 4 | 5 | 6 | 5 | 5 | 4 | 6 | 5 | 4 | Medium |
| P223 | 33 | 1 | 5 | 7 | 7 | 7 | 7 | 6 | 7 | 7 | 4 | 8 | 7 | 7 | 4 | 4 | 5 | 5 | 5 | 5 | 4 | 6 | 5 | High |
| P224 | 28 | 2 | 1 | 6 | 7 | 5 | 3 | 2 | 6 | 2 | 3 | 3 | 2 | 2 | 3 | 3 | 7 | 7 | 4 | 8 | 7 | 7 | 5 | Medium |

**FIGURE 13.3**
Portion of cancer patients' dataset used in this study.

It can be inferred from above that up to $\varepsilon$ value less than 0.2 the loss in user privacy is limited to 3% but more than the value of 0.2 for $\varepsilon$, which increases the degree of privacy loss which may be unacceptable for many users in a cyber-physical ecosystem. The graph in Figure 13.6 is plotted between Epsilon value and the percentage of data utility.

It can be easily inferred that as the value of $\varepsilon$ increases the data utility reduces substantially. It can be observed from the plot above that there is a significant drop in data utility value when $\varepsilon$ value increases from 0.1 to 0.3. For most practical purposes, this is not desired; e.g., in highly dynamic systems like healthcare CPS or smart cities or smart industrial environments. The data utility is desired trade-off with privacy. The data sensitivity value can also be regarded as an important factor when selecting the $\varepsilon$ values. In most of the datasets used in CPS, sensitivity is considered in the range of (0, 1). The values can be

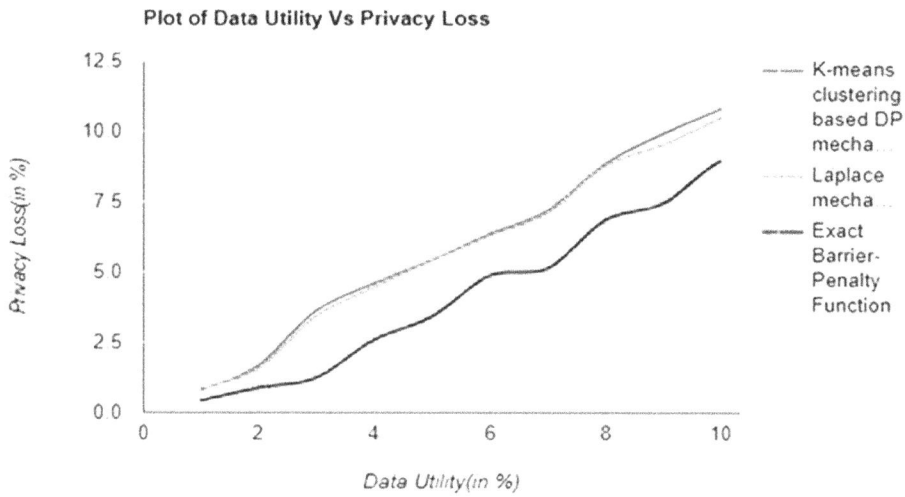

**FIGURE 13.4**
Comparative performance between most promising privacy preservation mechanisms in cyber-physical systems.

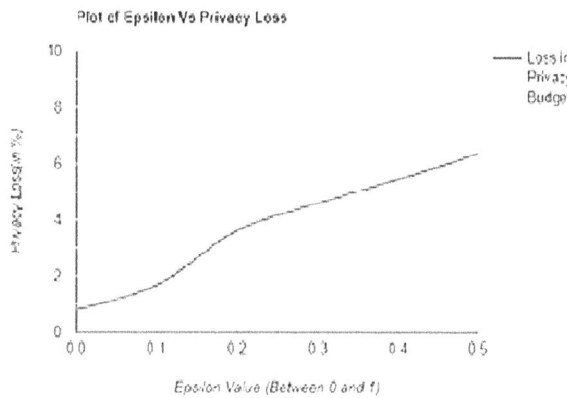

**FIGURE 13.5**
Loss of privacy for selected Epsilon values.

**TABLE 13.2**

Age Ranges of Cancer Patients Based on Dataset Obtained from Kaggle with the Application of Exact Barrier-Penalty Function Applied at $\varepsilon = 0.11$

| Age Range of Patients | Number of Cancer Patients (real value) | Number of Cancer Patients (value obtained after application of exact barrier-penalty function mechanism) |
| --- | --- | --- |
| 15–25 | 166 | 94 |
| 26–35 | 333 | 80 |
| 36–45 | 277 | 128 |
| 46–55 | 157 | 14 |
| 56–65 | 62 | 219 |
| 66–75 | 11 | 27 |

**TABLE 13.3**

Results of Privacy Loss for Different Ɛ Values Used in Minimization of the Objective Function

| No. of Iterations | Ɛ Value | Privacy Loss (in %) |
| --- | --- | --- |
| 10 | 0.11 | 14.68 |
| 10 | 0.20 | 19.88 |
| 10 | 0.30 | 24.03 |
| 10 | 0.50 | 31.51 |

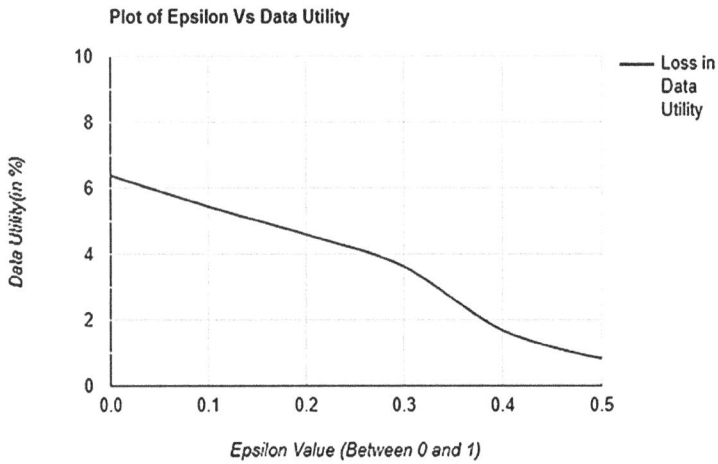

**FIGURE 13.6**
Loss of data utility for selected Epsilon values.

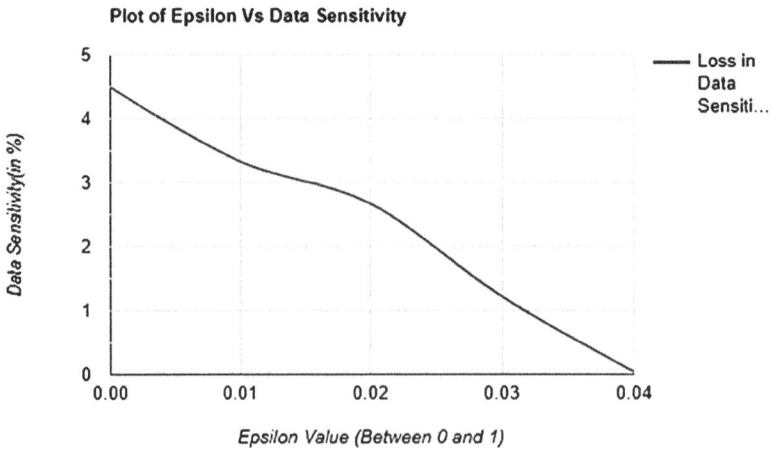

**FIGURE 13.7**
Loss of data sensitivity for selected Epsilon values.

adjusted based on composition theorems and a few auxiliary features. These auxiliary features can hide the sensitivity values remarkably. Proceeding further, Figure 13.7 shows the plot between $\varepsilon$ values and degree of data sensitivity.

It can be observed from the above graph that on a scale of 0–5% data sensitivity range, the $\varepsilon$ values can be selected from anywhere between 0.00 and 0.02. Once it crosses the range of 0.02, the measure of data sensitivity also gets minimized rapidly. Once the data sensitivity is out of control then the penalty function crosses the barrier and the privacy loss becomes undesirable. We can conclude that privacy preservation is possible only when the data sensitivity values are in control.

## 13.6 Future Scope

Much of the success of our proposed mechanism depends on the technique used to solve the intermediate problems, which in turn depends on the complexity of the CPS model. One thing that ought to be done preceding endeavoring to address a nonlinear program utilizing a penalty function technique is proportional to the constraints, so the penalty produced by each is about a similar magnitude. This scaling activity is expected to guarantee that no subset of the constraints impacts the pursuit interaction. On the off chance that a few constraints are prevailing, the calculation will control toward an answer that fulfills those constraints to the detriment of looking for the base. In a like way, the underlying worth of the penalty boundary ought to be fixed with the goal that the magnitude of the penalty term isn't a lot more modest than the magnitude of the objective function. In the event that an irregularity exists, the impact of the objective function could guide the calculation to head toward an unbounded least even within the sight of unsatisfied constraints. Regardless, convergence might be incredibly lethargic.

## 13.7 Conclusion

Privacy issues are very critical for the CPS ecosystem because people are becoming more and more aware of how their personal data is being used and monitored in the public domain. The very success of CPS functionality depends on the minimum amount of privacy loss of the users. Agreed that many CPS designers still don't take privacy as a quality while designing a robust CPS architecture, the rate of success of a CPS will surely depend on whether the CPS becomes even more trustworthy. In this direction, an intuitive privacy preservation scheme must be put into practice. The CPSs have to face different privacy attacks to access basic information or records from public or private datasets by malicious attackers. Among many mechanisms, one ideal answer to beat this challenge is, protection perils is preserving information by noise expansion utilizing differential security perturbation-based algorithms. In this chapter, a novel approach to identify the limit of privacy loss for an entity in the CPS ecosystem so that the user may feel no threat to their privacy while operating in the concerned environment is advocated. A comprehensive cover for the crucial dimensions and aspects of differential privacy with preservation of private and other sensitive data during system implementations in major CPSs areas is being provided by this paper. Penalty and barrier methods are among the most powerful class of algorithms available for tackling general nonlinear advancement issues. This statement is supported by the way that these techniques will converge to something like a nearby least as a rule, regardless of the convexity characteristics of the objective function and constraints. The proposed approach reduces the loss of privacy to a substantial degree and maintains a fair degree of data utility also. This mechanism will help to hide sensitive details of a user in the CPS ecosystem.

---

## References

Bhaskar, R., Luyao, N., Andrew, C., Linda, B., and Radha, P. (2020). Privacy-Preserving Resilience of Cyber-Physical Systems to Adversaries. *59th IEEE Conference on Decision and Control (CDC)*, pp. 3785–3792.

Bi, M., Wang, Y., Cai, Z., and Tong, X. (2020). A privacy-preserving mechanism based on local differential privacy in edge computing. *China Communications* vol. 17, pp. 50–65.

Degue, K. H. (2021). *Secure and Privacy-Preserving Cyber-Physical Systems: Phdthesis.* Polytechnique Montreal.

Desai, S., Alhadad, R., Chilamkurti, N., and Mahmood, A.. (2018). A survey of privacy preserving schemes in IoE enabled Smart Grid Advanced Metering Infrastructure. *Cluster Computing* vo. 22, pp. 1–27.

Farokhi & Farhad. (2020). Temporally discounted differential privacy for evolving datasets on an infinite horizon. *ACM/IEEE 11th International Conference on Cyber-Physical Systems (ICCPS)*, pp. 1–8.

Ghayyur, S., Chen, Y., Yus, R., Machanavajjhala, A., Hay, M., Miklau, G., and Mehrotra, S. (2018). IOT-detective: Analyzing IOT data under differential privacy. *Proceedings of International Conference on Management of Data*. ACM. pp. 1725–1728.

GitHub. (2020) [online]. Available at: https://github.com/anonymized

Heng, Y., Jiqiang, L., Wei, W., Ping, L., and Li, J. (2020). Secure and efficient outsourcing differential privacy data release scheme in Cyber–physical system. *Future Generation Computer Systems* vol. 108, pp. 1314–1323.

Hou, Jun, Li, Q., Cui, S., Meng, S., Zhang, S., Ni, Z., and Tian, Y. (2020). Low-cohesion differential privacy protection for industrial internet. *The Journal of Supercomputing* vol. 76, pp. 8450–8472.

Jairo, G., Alvaro, C., Murat, K., and Jonathan, K.. (2020). Adversarial classification under differential privacy. *Network and Distributed Systems Security (NDSS) Symposium*.

Jia, R., Dong, R., Sastry, S. S., and Sapnos, C. J. (2017). Privacy-enhanced architecture for occupancy-based hvac control. *ACM/IEEE 8th International Conference on Cyber-Physical Systems (ICCPS)*, pp. 177–186.

Jian, X., Laiwen, W., Wei, W., Andi, W., Yu, Z., and Fucai, Z. (2020). Privacy-preserving data integrity verification by using lightweight streaming authenticated data structures for healthcare cyber–physical system. *Future Generation Computer Systems* vol. 108. pp. 1287–1296.

Jiang, B., Li, J., Yue, G., and Song, H. (2021). Differential privacy for industrial internet of things: opportunities, applications and challenges. *IEEE Internet of Things Journal* vol. 8, pp. 10430–10451.

Keshk, M., Sitnikova, E., Moustafa, N., Hu, J., and Khalil, I. (2019). An integrated framework for privacy-preserving based anomaly detection for cyber-physical systems. *IEEE Transactions on Sustainable Computing*, vol. 6, pp. 66–79.

Qu, Y., Yu, S., Zhou, W., Chen, S., and Wu, J. (2020). Customizable reliable privacy-preserving data sharing in cyber-physical social networks. *IEEE Transactions on Network Science and Engineering* vol. 8, pp. 269–281.

# *Index*

For Product Safety Concerns and Information please contact our EU
representative GPSR@taylorandfrancis.com
Taylor & Francis Verlag GmbH, Kaufingerstraße 24, 80331 München, Germany